KEYWORDS IN MAKING

KEYWORDS IN MAKING

A RHETORICAL PRIMER

Edited by Jason Tham

Parlor Press
Anderson, South Carolina
www.parlorpress.com

Parlor Press LLC, Anderson, South Carolina, USA

© 2024 by Parlor Press
Printed in the United States of America on acid-free paper.

SAN: 254-8879

Library of Congress Cataloging-in-Publication Data on File

978-1-64317-480-8 (paperback)
978-1-64317-481-5 (pdf)
978-1-64317-482-2 (epub)

2 3 4 5

Cover image: Photo by Kier in Sight on Unsplash. Used by permission.
Book Design by David Blakesley
Copyediting by Fran Chapman

Parlor Press, LLC is an independent publisher of scholarly and trade titles in print
and multimedia formats. This book is available in paper and ebook formats from
Parlor Press on the World Wide Web at https://www.parlorpress.com or through
online and brick-and-mortar bookstores. For submission information or to find
out about Parlor Press publications, write to Parlor Press, 3015 Brackenberry Drive,
Anderson, South Carolina, 29621, or email editor@parlorpress.com.

In Memoriam

Halcyon Lawrence, Johndan Johnson-Eilola,
and Bill Hart-Davidson

Contents

Acknowledgments   *xv*

Editor's Introduction   *xvii*
    Jason Tham

Part 1: Theoretical Concepts

1. Accessibility   5
    Cody A. Jackson and Justin H. Cook

2. Active Learning   10
    Joe Moses

3. Advocacy   13
    John Joseph Silvestro

4. Aesthetics   16
    Scott Sundvall

5. Ambiguity   20
    Amanda M. May

6. Assemblage   23
    Amanda M. May

7. Cognitive Dissonance   27
    Johansen Quijano

8. Constructionism   30
    Merideth Garcia

9. Context and Contextual Design   33
    Thomas M. Geary

10. Creative Confidence   36
    Kamila Albert

11. Cultural Intelligence   39
    Nitya Pandey

12. Curiosity   42
    Alexandra Catá-Ross

13. Digital Rhetoric   *45*
    Kevin Brock

14. Electracy   *48*
    Kristopher Purzycki

15. Embodiment   *51*
    Heather Listhartke

16. Environments   *55*
    Jacob Craig

17. Ethics   *58*
    Steve Holmes

18. Heuristics   *62*
    Derek N. Mueller

19. Inter/Cross-Disciplinarity   *65*
    Ann Shivers-McNair

20. Invention   *68*
    Erin Kathleen Bahl

21. Learning Diversity   *71*
    Michael Riendeau

22. Local   *75*
    Madison Jones

23. Maker Culture   *80*
    Johansen Quijano

24. Maker Movement   *83*
    Johansen Quijano

25. Materiality   *86*
    Jason Tham

26. Memory   *90*
    Joe Cirio

27. Multigenre   *93*
    Megan Marshall

Contents

28. Perseverance   96
    Megan Poole

29. Play/Playful   99
    A. Nicole Pfannenstiel

30. Postdigital Aesthetics   102
    Jialei Jiang

31. Posthuman Practice   105
    Jialei Jiang

32. Radical Imagination   109
    Stephanie West-Puckett

33. Social Constructivism   112
    Laura Roberts

34. Visual Rhetoric   115
    Jason Tham

35. Visual Semiotics   120
    Jason Tham

Part 2: Practices

36. Augment/Augmentation   125
    Jacob Greene

37. Bias Toward Action   128
    Katherine Goodman

38. Brand/Branding   130
    Scott Sundvall

39. Coding   133
    Charles Woods

40. Coding Literacy   136
    Antonio Byrd

41. Community Engagement   139
    Sweta Baniya

42. Community of Practice   141
    Jeff Naftzinger

x

43. Composition Commons   *144*
   Jess Clements

44. Creative Commons Licensing   *150*
   Quentin Vieregge

45. Curation   *153*
   Kathleen Blake Yancey

46. Data Visualization   *156*
   Liz Lane

47. Design Challenge & Makeathon   *159*
   Jason Tham

48. DIY/Do It Yourself   *162*
   Joy Santee

49. DIWO/Do It With Others   *165*
   Krys Gollihue

50. Extreme Situation   *168*
   R.J. Lambert

51. FabLab   *171*
   Estee Beck

52. Feminist Making   *174*
   Krys Gollihue

53. Growth vs. Fixed Mindsets   *177*
   Katherine Goodman

54. Hacking   *180*
   Sergio C. Figueiredo

55. Inclusive Design   *183*
   Kristen R. Moore

56. Information Design   *186*
   Sarah Welsh

57. Intellectual Property   *189*
   Heather Listhartke

Contents

58. Interaction Design   192
    Jennifer Sano-Franchini

59. Iterative Design   196
    Cody Reimer

60. Leverage Points   199
    Kristopher Purzycki

61. Maker Competencies   201
    Estee Beck

62. Maker Faire   206
    Quentin Vieregge

63. Makerspace   209
    Marijel (Maggie) Melo & Jason Tham

64. Mentoring   212
    Lyra Hilliard

65. Open Access   216
    Dana Lynn Driscoll

66. Physical Computing   219
    Emily F. Brooks

67. Project Management   222
    Sarah Young

68. Public Interest Design   225
    Max Renner

69. Radical Collaboration   228
    Joe Moses

70. Safety   231
    R.J. Lambert

71. Shared Leadership   234
    Ann Hill Duin

72. Social Innovation   238
    Nupoor Ranade

73. STEM & STEAM   242
    Mary E. Caulfield

74. Technical Communication   244
    Sara Doan

75. Tools   247
    Jacob Craig

76. Universal Design   250
    Ada Hubrig

77. User Experience   253
    Dennis Cheatham

78. User Requirements   255
    Joseph Bartolotta

79. Writing Studio   257
    Matthew Kim & JongHun Kim

Part 3: Methods

80. 3D Printing   263
    Charles Woods

81. A/B Testing   266
    Halcyon M. Lawrence

82. Affinity Diagramming   269
    Arthur Berger

83. Autoethnography   273
    Erin Kathleen Bahl

84. Bodystorming   276
    Chloe Anna Milligan

85. Card Sorting   279
    Joseph Bartolotta

86. Case Study   281
    Mandy Olejnik

Contents

87. Diary Study   203
    Elin Björling

88. Ethnography   286
    Renee Ann Drouin

89. Fieldwork   289
    Casey Boyle

90. Interviewing   292
    Ashanka Kumari

91. Journey Mapping   295
    Joe Moses

92. Peer Response   299
    Heather Listhartke

93. Photovoice   302
    Erin Brock Carlson

94. Postmortem   306
    Kristopher Purzycki

95. Reflection   308
    Kathleen Blake Yancey

96. Repair   312
    Thomas Karches

97. Teardown   315
    Jason Markins

98. Tinkering   318
    Danielle Koupf

99. User stories   321
    Ann Shivers-McNair

100. Wireframing   324
     Joseph Bartolotta

Contributors   327

Acknowledgments

It has certainly felt like forever to get this book into the world. This project started when we all were in a very different place (read: pre-pandemic). Right as I was completing my doctoral degree in spring of 2019, I was inspired to create a crowdsourced material that could guide us in understanding what "making" means in the world of rhetoric, writing, and communication design. This motivation was in part due to the many questions I received from folks who have read or heard about my dissertation—What is it about "making" or "maker culture" that deserves our attention? What does it mean to make? What theories and concepts does making draw from? How does making fit in writing, communication, or rhetorical studies? In response, I thought a lexical collection might serve to address some of these questions. Hence this project!

To my surprise, the initial call for contribution to this project received an overwhelming response, so much so that it resulted in two separate collections—one on *design thinking* (published by The WAC Clearinghouse/University Press of Colorado), and one on *making* (this one). I thank the contributors to this collection, all seventy-five of them, for believing in this project and hanging on even when recent world events made it incredibly tough to do any scholarship such as this. (Writing, really, isn't a priority in many of our lives.) I am particularly grateful to David Blakesley of Parlor Press for shepherding us through the proposal and production process. I am also thankful to the anonymous reviewers who provided constructive feedback that has strengthened this collection. Thanks to Fran Chapman for her careful copyediting.

I would like to acknowledge my colleagues and mentors who have shown me ways to perform research. My university and department administrators have been kind and supportive of my scholarly pursuits, lending to projects that wouldn't have been possible without their trust. This collection is an example. And as always, my students are the source of my devotion to academia. I have worked to produce this project so it can be a resource to students in communication and writing who identify as makers. My hope is that it serves them in learning the complexity of making, and empowering them to advocate for other makers.

Writing, like making, is seldom a lonely activity. Throughout this project, I have my partner, Kamm, to thank for caring and understanding my writing routines (thank you for helping me make space to write and revise). And I also have my furry babies, Cornelius, Malia, and Athena, to praise for their unconditional affection whether or not I was going to get published; they demand only treats and cuddles.

Editor's Introduction

Jason Tham

*K*eywords in Making: A Rhetorical Primer yields a multidisciplinary response to the "design turn" in writing studies and the humanities writ large. The genesis of this project was due to the inconsistent references found in the literature and teaching of design in writing studies, specifically rhetoric and technical communication. When I presented research on design at conferences, local and national alike, scholars from across the field have recommended (vastly) different theoretical directions for my work. Those who identify as writing and literacy scholars often point me to semiotics and education theories; rhetoricians and compositionists would cite visual and digital rhetorics; and my colleagues in technical communication and user experience research have emphasized human-centered design principles and collaborative methods. While there are certainly overlapping conversations around design-centric, human-oriented approaches in pedagogy and practice, I recognize this moment as an opportunity for a shared resource for those interested in these approaches. This collection is informed by this exigence, created to provide a common lexicon around design activities, along with the goal of inviting various perspectives to which particular keywords are conceptualized, applied, and studied. However, just as it's alluded through the cover art of this book, the words or terms included in this collection are entangled with forces of their use, teaching, and application that could influence the way they are consumed. So, reader's discretion is advised!

Like many keyword projects across disciplines, this collection was inspired by Raymond Williams's (1976) *Keywords: A Vocabulary of Culture and Society* (1976), which showed the importance of examining the language that illuminates any given field. Some similar contemporary keyword collections include language and literacy (Carter, 1995), literature (Nel, Paul, & Christensen, 2021), creative writing (Bishop & Starkey, 2006), biology (Fox Keller & Lloyd), cultural studies (Burgett & Hendler, 2020; Edwards, Ferguson, & Ogbar, 2018; Schlund-Vials, Wong, & Võ, 2015; Vargas, La Fountain-Stokes, & Mirabal, 2017), disability studies (Adams, Reiss, & Serlin, 2015), journalism and media studies (Ouellette & Gray, 2017; Zelizer & Allan, 2010), sound studies (Novak & Sakakeeny, 2015), environmental studies (Adamson, Gleason, & Pellow, 2016), travel writing (Forsdick, Kinsley,

& Walchester, 2019), gender and sexuality studies (The Keywords Feminist Editorial Collective, 2021), rhetoric (Lanham, 1991), composition and writing studies (Heilker & Vandenberg, 1996, 2015), and technical writing and communication (Gallon, 2016; Yu & Buehl, 2023).

Each of these volumes presents the principal ideas in their respective knowledge fields. Over time, they become the historical accounts of a field through its evolving lexicon. For example, Paul Heilker and Peter Vandenberg's *Keywords in Writing Studies* (2015), a sequel to their *Keywords in Composition Studies* (1996), has traced the shifting theoretical, educational, professional, and institutional developments across a span of two decades. In updating Williams's collection, Colin MacCabe and Holly Yanacek (2018) argued that documenting keywords represents a necessary effort any field should attempt with diligence. Such a scholarly exercise makes the case for awareness of the ways in which significant words shape and inform our inquiries.

I started this project with the intention of creating a primer for researchers and practitioners interested in "making" as an emerging approach in the classroom as well as the workplace. The entries included in this collection aim to give readers clarity in popular concepts related to these approaches through contextual definitions and some short case scenarios. The collection was not strictly designed for scholars in writing studies, nor was it meant to be an enterprise dictionary. It is edited for a wider audience to whom picking up this collection means engaging with a diverse group of makers. The contributors to this project include college professors, graduate students, a few industry practitioners, and high school teachers—all of whom share a common passion for meaningful making, one way or another. This collection will make a handy reference for students, graduate and undergraduate alike, who are beginning their quests in applying or practicing making as part of their learning. Instructors may assign this collection in writing studies seminars where students need quick access to key concepts and examples for their coursework or scholarly projects.

Writing studies as a field is increasingly invested in design-centric approaches in pedagogy and practice. Among the first to introduce us to design as a potential paradigm shift in the way we perceive writing is Charles Kostelnick, whose landmark article in *College Composition and Communication*, "Process Paradigm in Design and Composition: Affinities and Directions" (1989) critiqued the then buzzword, "process pedagogy," and offered design as an augmentation to the writing process. Twenty years later, Richard Marback (2009) again offered design thinking as a "new" paradigm for writing studies. Given the increased attention to new media, writing studies as a whole is becoming more accepting of design approaches to composing, especially when it involves multimedia technology and "wicked" communicative problems that require solutions beyond text-only mediation. Scholars like James Purdy (2014) and Carrie Leverenz (2014) have argued for design as a generative as well as productive methodology for teaching writing and professional communication. "Invoking design," Purdy stated, "can serve to answer Jody Shipka's call for the discipline to focus on all communicative practices, not just writing" (2014, p. 73).

Enjoying a unique limelight in current writing studies literature is "making," a notion popularized by a social initiative called the Maker Movement. Following the 2013 Computers and Writing conference keynote by James Paul Gee, making as a social practice continues to manifest in conference programs, journal articles, and books. Chet Breaux (2017) has observed that many writing scholars are already very interested in the practices used by "makers" and the artifacts they create. In her 2016 Conference on College Composition and Communication (CCCC) Chair's Address, "Making. Disrupting. Innovating," Joyce Locke Carter called on writing studies scholars to value making as a valid and plausible way of learning anything in the twenty-first century, including writing. Carter's exigence was built upon the historical impact that making has on our field and its advancement. In her speech, Carter called our attention to several innovative instances, such as Daedalus (University of Texas), Eli Review (Michigan State University), BABEL (MIT), and EyeGuide (Texas Tech University), all of which have helped define writing studies as a productive discipline that contributes to the betterment of our knowledge in society (Carter, 2016).

In addition to Carter's motivation, I subscribe to a characterization of our field that Johndan Johnson-Eilola and Stuart Selber (2013) set out for technical communication, which makes our field an eminent leader in theorizing making practices:

> Technical communicators do not merely learn skills; they must also learn how to learn new skills, upgrading and augmenting their abilities as they mature in careers, analyzing the matches and mismatches between what they currently know and what a communication situation demands . . . [They] must learn to become reflective problem solvers. (p. 3)

Johnson-Eilola and Selber consider problem-solving as a productive characterization for it acknowledges the extent to which our field contributes to technological development and its use, the interpretation of rhetorical situations, and the design of viable solutions based on context, complexity of the tasks and their characteristics. Making, as an intentional problem-solving activity, invites a certain kind of thinking and responding that is important to writers and designers today. The entrepreneurial nature of the current maker culture provides a vital foundation for students to work in a non-linear process, trying multiple strategies to arrive at a plausible solution. It encourages students to practice employing multiple modalities to construct their solutions. Thus, it creates new approaches for teaching and learning that respond to existing industry needs. This collection provides the necessary vocabulary for understanding and activating these approaches.

To achieve the purpose of a reference guide and not an encyclopedia, I have asked contributors to compose their entries succinctly. Every keyword is explained in five hundred to one thousand words, making it easy for someone who is just looking to get an orientation to the keyword in the context of making and design. Each entry in this collection begins with a definition to introduce the featured keyword, followed by a synthesis of relevant research, sometimes coupled with fitting experiences of the contributor or a short case scenario to contextualize the keyword. It includes cross-references ("see also") to other entries within the collection that discuss similar subjects. Each entry also ends with five or more recommended resources provided by the contributor and me. To help readers navigate the collection, I have organized the entries into three structural categories:

- **Theoretical concepts**: Frameworks and perspectives that guide making.
- **Practices**: Ways of making across multiple contexts.
- **Methods**: Specific tools and techniques for making.

While these categories help organize the collection in a demi-logical sense, many entries fit in more than one of the categories due to their expansive nature. I encourage readers to embrace such ambiguity and avail themselves of a reading experience that is not strictly bound by these categories.

I should also note here that this project was initially conceptualized as a compendium of keywords in making *and* design thinking, but it has ultimately evolved into two separate collections, with the design thinking collection being published with The WAC Clearinghouse/University Press of Colorado (Tham, 2022). There is no repetition in the keywords included in both collections, so I encourage readers to also check out the design thinking collection for a more comprehensive learning about design and making.

Finally, modeled after innovative works like Guy McHenry's *Key Concepts in Surveillance Studies* (2017) and Cheryl Ball and Drew Loewe's *Bad Ideas about Writing* (2017), this collection is designed to be a shareable and reusable resource. Readers may reuse portions or all of this collection with attribution to the original texts and authors. The goal is for this collection to serve as a springboard for those deploying making in their work to achieve desirable outcomes.

References

Adams, R., Reiss, B., & Serlin, D. (Eds.). (2015). *Keywords for disability studies*. NYU Press.

Adamson, J., Gleason, W. A., & Pellow, D. N. (Eds.). (2016). *Keywords for environmental studies*. NYU Press.

Ball, C., & Loewe, D. (Ed.). (2017). *Bad ideas about writing*. West Virginia University Libraries/Digital Publishing Institute.

Bay, J., Johnson-Sheehan, R., & Cook, D. (2018). Design thinking via experiential learning: Thinking like an entrepreneur in technical communication courses. *Programmatic Perspectives, 10*(1), 172–200.

Bishop, W., & Starkey, D. (2006). *Keywords in creative writing*. Utah State University Press/University Press of Colorado.

Breaux, C. (2017). Why making? *Computers and Composition, 44*, 27–35. https://doi.org/10.1016/j.compcom.2017.03.005

Burgett, B., & Hendler, G. (Eds.). (2020). *Keywords for American cultural studies* (3rd ed.). NYU Press.

Carter, J. (2016). Making. Disrupting. Innovating. *College Composition and Communication, 68*(2), 378–408. https://www.jstor.org/stable/44783568

Carter, R. (1995). *Keywords in language and literacy*. Routledge.

Edwards, E. R., Ferguson, R. A., & Ogbar, J. O. G. (Eds.). (2018). *Keywords for African American studies*. NYU Press.

Forsdick, C., Kinsley, Z., & Walchester, K. (Eds.). (2019). *Keywords in travel writing studies: A critical glossary*. Anthem Press.

Fox Keller, E., & Lloyd, E. A. (Eds.). (1992). *Keywords in evolutionary biology*. Harvard University Press.

Gallon, R. (Eds.). (2016). *The language of technical communication*. XML Press.

Gee. J.P. (2013, June). *Writing in the age of the maker movement*. Keynote presented at 2013 Computers and Writing conference, Frostburg, MD.

Heilker, P., & Vandenberg, P. (Eds.). (1996). *Keywords in composition studies*. Boynton/Cook Publishers.

Heilker, P., & Vandenberg, P. (Eds.). (2015). *Keywords in writing studies*. Utah State University Press/University Press of Colorado.

Johnson-Eilola, J., & Selber, S. (Eds.). (2013). *Solving problems in technical communication*. University of Chicago Press.

Kostelnick, C. (1989). Process paradigm in design and composition: Affinities and directions. *College Composition and Communication, 40*(3), 267–281. https://www.jstor.org/stable/357774

Lanham, R. A. (1991). *A handlist of rhetorical terms* (2nd ed.). University of California Press.

Leverenz, C. (2014). Design thinking and the wicked problem of teaching writing. *Computers and Composition, 33*, 1–12. https://doi.org/10.1016/j.compcom.2014.07.001

Marback, R. (2009). Embracing wicked problems: The turn to design in composition studies. *College Composition and Communication, 61*, 397–419. https://www.jstor.org/stable/40593465

MacCabe, C., & Yanacek, H. (Eds.): (2018). *Keywords for today: A 21–st century vocabulary*. Oxford University Press.

McHenry, G. (Ed.). (2017). *Key concepts in surveillance studies*. https://surveillancestudies.pressbooks.com/

Nel, P., Paul, L., & Christensen, N. (Eds.). (2021). *Keywords for children's literature* (2nd ed.). NYU Press.

Novak, D., & Sakakeeny, M. (Eds.). (2015). *Keywords in sound*. Duke University Press.

Ouellette, L., & Gray, J. (Eds.). (2017). *Keywords for media studies*. NYU Press.

Purdy, J. (2014). What can design thinking offer writing studies? *College Composition and Communication, 65*(4), 612–641. https://www.jstor.org/stable/43490875

Schlund-Vials, C. J., Wong, K. S., & Vò, L. T. (Eds.). (2015). *Keywords for Asian American studies*. NYU Press.

Tham, J. (Ed.). (2022). *Keywords in design thinking: A lexical primer for technical communicators and designers*. The WAC Clearinghouse/University Press of Colorado. https://wac.colostate.edu/books/tpc/design/

The Keywords Feminist Editorial Collective. (Ed.). (2021). *Keywords for gender and sexuality studies*. NYU Press.

Vargas, D. R., La Fountain-Stokes, L., & Mirabal, N. R. (Eds.). (2017). *Keywords for Latina/o studies*. NYU Press.

Williams, R. (1976). *Keywords: A vocabulary of culture and society*. Croom Helm.

Yu, H., & Buehl, J. (Eds.). (2023). *Keywords in technical and professional communication*. The WAC Clearinghouse/University Press of Colorado. https://wac.colostate.edu/books/tpc/tpc/

Zelizer, B., & Allan, S. (2010). *Keywords in news and journalism studies*. McGraw Hill/Open University Press.

KEYWORDS IN MAKING

PART 1:
THEORETICAL CONCEPTS

1. Accessibility

Cody A. Jackson and Justin H. Cook

Accessibility is the practice of making things, processes, and environments possible, sensible, and meaningful for all people rather than a privileged population of users.

For inventors and technical communicators, accessibility should be a priority in design. This means maximizing everyone's chances of engaging with a designed product and reducing undesirable encounters due to limited access. Access to products or activities can be impeded by sociocultural and political barriers that unevenly impact disabled and neurodivergent people, particularly along lines of:

- Cognitive, physical mobility, auditory, verbal, or ocular disabilities
- Age, language, culture, education, and other personal characteristics
- Economic position, class status, living arrangements
- Access to technology and resources

For designers, accessibility can be about compliance with the rules, regulations, or laws of accessibility set forth by organizations, agencies, or communities. Two commonly referred to compliance standards are Section 508 of the Rehabilitation Act of 1973 and the Web Content Accessibility Guidelines (WCAG).

Section 508 is a US governmental regulation that covers disability policies and accessibility compliance requirements for government entities, federal employers, and subcontractors and their information and communication technologies. Under Section 508, agencies must give disabled employees and members of the public access to information comparable to the access available to others.

The Web Accessibility Initiative (WAI) is an international organization born of the World Wide Web Consortium (W3C) that recommended the accessibility guidelines presented in WCAG. Though primarily focused on accessibility of content on the web, these standards inform accessibility requirements for content in other contexts (e.g., screen-based documents, PDFs, etc.).

Compliance should not be the only reason for practitioners to consider accessibility a requirement in design. Accessibility is an ethical decision designers and communicators actively make to demonstrate care and user advocacy. Accessibility is contextual. It is a cultural practice that strives to achieve equity for users.

In this keyword collection on making, we hope to gesture toward ways that design can be expanded for and through anti-ableist advocacies. However, there is no accessibility checklist commensurate with disability justice (Wood et al., 2014; Oswal & Melonçon, 2017). Our ongoing commitment to one another is found within the mundane grooves of everyday life (Hamraie, 2016). Accessibility is a project that is always forward-dawning, always on the horizon of collective possibility and responsibility (Segarra, 2017). But accessibility can't be relegated to the work of the future; rather, it must be a conscious undoing of ableist frameworks that individualize the project of access, a project that will never be anything but a *relational* (see Licona et al., 2015), *collective* struggle for intersectional methodologies of disability justice.

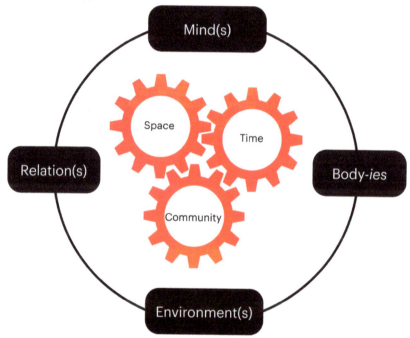

Fig. 1.1. A possible model for accessibility and design. Source: Contributing authors.

Fig. 1.1. is a possible framework we propose for accessible design. It shows two static images at play. The first, on the outside, are four interconnected, non-directional squares that are not intended to privilege one over the other. Named clockwise they include mind(s), relation(s), environment(s), and body-ies. Inside of this shape are three interconnected mechanical gears with curved arrows that signify their relations with one another. Starting with the top gear and ending with the bottom gear: space, time, and community.

This framework is one of many ways that accessibility can be addressed and, more importantly, practiced. For example, when considering spatialization, course design should address not only the physical accessibility of the classroom but the accessibility of the course's digital components such as the learning management platform, and digital communication expectations and practices. Second, course design must necessarily understand the ways that bodyminds navigate courses and course assignments. As time and space are experienced differently depending on our own bodymind realities, these divergent experiences must be accounted for when considering project deadlines, project requirements, and course expectations. Finally, we acknowledge that this model is necessarily malleable depending on the context of the course and community.

The collective work toward accessibility *as* disability justice should be oriented *toward* anti-institutional modes of relationality (Crawley, 2018) and *away* from individualized notions of responsibility, care, and action. We gesture toward *accessibility as a way of life*. We have developed a heuristic for continuing these conversations (see Fig. 1.1). Working with Liz Jackson's (2019) recent work, the model seeks to convey a methodology of *design questioning*, as well as techniques of *design thinking–feeling*, that reorient design as necessarily disruptive, relational, and ongoing. Among many questions, we urge designers to consider:

- How are we interrogating our use of "we" in conversations about accessibility? How can designers, teachers, and higher education professionals be mindful of the political weight of the invoked "we" in these conversations?
- How are *we* designing spaces that are flexible and shape-shifting: spaces that are open to the excesses of disabled, neurodivergent, neuroqueer (Yergeau, 2017) bodyminds (Price, 2011)?
- How are *we* expanding notions of access to make room for conversations about disability to necessarily intersect with and through conversations about race, gender, sexuality, gender expression, gender identity, and nationality?
- How are *we* designing textual spaces that make room for screen-reader technology, divergent cognitive processing, predictable spatial environments for blind or vision-impaired readers, and accessible captioning (Butler, 2018) platforms?
- How are *we* designing spaces and environments that are aware of Multiple Chemical Sensitivity (MCR)?

- How are *we* designing curricula that foreground work written by *disabled* disability studies scholars, activists, and writers?
- How are *we* cultivating activist spaces and groups within our respective institutions that are led *by* and run *for* disabled students, faculty, and staff?
- How are *we* working to ensure that access to our respective disability services offices is at the *front* of our campus buildings? Similarly, how are *we* working to reshape the terrain of university campuses so that they are flexible to disabled bodyminds (Dolmage, 2017)?
- How are *we* designing university functions and program curricula (Wood, 2017; Hitt, 2018) that are attuned to the myriad ways in which time is experienced and embodied?
- How are *we* working to build an academic job market (Sano-Franchini 2016) that is more accessible and aware of the uneven distribution of labor (Gaeta, 2019) placed onto marginalized bodyminds?
- How are *we* reimagining publishing and conferencing protocols that make space for disabled, poor, and marginalized bodyminds in ways beyond the text? In other words, how are *we* embodying our commitment to access in everyday interactions with others?

More importantly, *what has been left out here*? How does accessibility, in and beyond the stakes laid out here, make possible *and* impossible the very formation of the communities we aspire to work within, for, and alongside?

References

Butler, J. (2018). Integral captions and subtitles: Designing a space for embodied rhetorics and visual access. *Rhetoric Review*, *37*(3), 286–299.

Crawley, A. (2018, May 19). Ghosts. The New Inquiry. https://thenewinquiry.com/ghosts/.

Dolmage, J. (2017). *Academic ableism: Disability and higher education*. University of Michigan Press.

Gaeta, A. (2019, June 3). Cripping emotional labor: A field guide. Disability Visibility Project. https://disabilityvisibilityproject.com/2019/06/03/cripping-emotional-labor-a-field-guide

Hamraie, A. (2016). Beyond accommodation: Disability, feminist philosophy, and the design of everyday academic life. *philoSOPHIA*, *6*(2), 259–271.

Hitt, A. (2018). Foregrounding accessibility through (inclusive) universal design in professional communication curricula. *Business and Professional Communication Quarterly*, *81*(1), 52–65.

Jackson, L. (2019, May 28). *Email correspondence*. Design questioning. Retrieved from https://www.disabledlist.org/.

Licona, A. C., & Chávez, K. R. (2015). Relational literacies and their coalitional possibilities. *Peitho Journal, 18*(1), 96–107.

Oswal, S. K., & Melonçon, L. (2017). Saying no to the checklist: Shifting an ideology of normalcy to an ideology of inclusion in Online Writing Instruction. *WPA: Writing Program Administration, 40*(3), 61–77.

Price, M. (2011). *Mad at school: Rhetorics of mental disability and academic life*. University of Michigan Press.

Sano-Franchini, J. (2016). "It's like writing yourself into a codependent relationship with someone who doesn't even want you!": Emotional labor, intimacy, and the academic job market in rhetoric and composition. *College Composition and Communication, 68*(1), 98–124.

Segarra, A. (2017, May 30). Interview: Annie Segarra. The future is accessible. *bonfire*. https://blog.bonfire.com/interview-annie-segarra-future-accessible/.

Wood, T. (2017). Cripping time in the college composition classroom. *College Composition and Communication, 69*(2), 260–286.

Wood, T., Dolmage, J., Price, M., & Lewiecki-Wilson, C. (2014). Moving beyond disability 2.0 in Composition Studies. *Composition Studies, 42*(2), 147–150.

Yergeau, R. (2017). *Authoring autism: On rhetoric and neurological queerness*. Durham, NC: Duke University Press.

Recommended Resources

Bennett, K. C., & Hannah, M. A. (2022). Transforming the rights-based encounter: Disability rights, disability justice, and the ethics of access. *Journal of Business and Technical Communication, 36*(3), 326–354. DOI: 10.1177/10506519221087960

Berne, P., Morales, A. L., Langstaff, D., & Invalid, S. (2018). Ten principles of disability justice. *WSQ: Women's Studies Quarterly, 46*(1), 227–230.

Hamraie, A. (2018). Mapping access: Digital humanities, disability justice, and sociospatial practice. *American Quarterly, 70*(3), 455–482. DOI 10.1353/aq.2018.0031

McKinsey & Co. (2020). Accessible design means better design. https://www.mckinsey.com/capabilities/mckinsey-design/how-we-help-clients/design-blog/accessible-design-means-better-design

Niggl, K. (2021). *A guide to accessible design for connected products and services*. Futurice. http://bit.ly/katja-niggl

See also: Advocacy, Inclusive Design, Learning Diversity, Embodiment, Universal Design

2. Active Learning

Joe Moses

Many consider design and making to be active ways to engage learners. Active learning refers to activities in which students engage in doing work and then think about what they did (Bonwell & Eison, 1991). How do learners learn by doing? They learn by interacting with each other while defining problems, testing ideas, drafting content or prototyping solutions, discussing outcomes, and revising or reworking solutions based on feedback. Researchers commonly contrast active learning with passive learning, which is characterized by students sitting still while listening to a lecture, and they attribute a variety of advantages to the active approach. In addition to the influential work of Charles C. Bonwell and James A. Eison mentioned above, researchers have identified significant improvement in learning outcomes in science, technology, engineering and math (STEM) courses when active learning complements lectures (Freeman et al., 2014).

Active learning engages participants in critical thinking, making comparisons, building prototypes, and applying deductive or inductive reasoning—all of which are activities that make demands on students that are different from the demands of writing notes or memorizing information for a test. Roots of contemporary thinking about active learning reach back to a distinction that philosopher Gilbert Ryle made between knowing how and knowing that. Ryle argues that learning is a demonstration of thought applied to action (1949, p. 41).

Knowing how	Knowing that
3D printing	Listening to a lecture
Coding	Writing down lecture notes
Laser cutting	Watching a video
Sewing	
Soldering	
Visualizing data	
Writing	

Fig. 2.1. Active learning designs enable students to apply information, principles, concepts, or standards (knowing that) to tasks performed in pursuit of a goal such as problem solving (knowing how).

Design thinking as an active learning practice: With empathy as a guiding principle of problem solving in design thinking, practitioners seek empathy with consumers, product designers, developers, investors, policymakers, community organizers, and teammates—anyone with a stake in the outcome of problem-solving activities.

Active learning through problem definition: Good problem solvers know that in order to create effective solutions they must get to the roots of the issue. Design thinking asks participants to articulate clear, actionable problem statements instead of pursuing solutions to what may otherwise be poorly defined problems.

Active learning through ideation: Design thinking offers a structure for ideation—or activities for making thinking visible. Rooted in human-centered design, design thinking draws from philosophies of accessibility, equity and social justice, user advocacy, and radical collaboration to inform change-making. Therefore, active learning invites thinking that originates from diverse perspectives and making thinking visible.

Active learning by prototyping: Among the most important distinctions of design thinking is the materializing of ideas through low-fi or rapid prototyping. The goal of prototyping is to create early, low-stakes, small-scale versions of ideas for testing. This human-centered process benefits designers, users—as many internal and external stakeholders as you can get into the same room—by giving dimension to ideas and inviting responses to the ideas from others.

Active learning from testing: Testing is less about seeking affirmation for a complete solution and more about learning how others interact and respond to a proposed direction toward a solution. Therefore, testing in active learning is frequent and routine rather than rare and always high-stakes.

Active learning through retrospection and iteration: Upon returning from testing sessions, design thinking practitioners often spend a good deal of time reviewing participant feedback and concerns. As an iterative problem-solving process, design thinking encourages practitioners to go back to their original drawing boards and revisit previously unconsidered problems. This retrospective approach instills in the practitioners a mindset that problem solving is an iterative process. Human problems are complex problems, and complex problems aren't easily solved in a single attempt.

References

Bonwell, C. C., & Eison, J. A. (1991). Active learning: Creating excitement in the classroom. *ASH#-ERIC Higher Education Report No. 1.* Washington, D.C.: The George Washington University, School of Education and Human Development.

Freeman, S., Eddy, S., McDonough, M., Smith, M., Okoroafor, N., Jordt, H., & Wenderoth, M. (2014). Active learning increases student performance in science, engineering, and mathematics. *Proceedings of the National Academy of Sciences of the United States of America, 111*(23), 8410–8415. http://www.jstor.org.ezp3.lib.umn.edu/stable/23776432

Ryle, G. (1949). *The concept of mind.* University of Chicago Press.

Recommended Resources

Abendroth, L. M. (2014). Design for social justice. Retrieved from https://www.du.edu/cme/media/documents/coleads2014-abendroth-powerpoint.pdf

Clark, R. M., Starbryla, L. M., & Gilbertson, L. M. (2018). Use of active learning and design thinking process to drive creative sustainable engineering design solutions. In *Proceedings of the 2018 ASEE Annual Conference & Exposition* (pp. 1–18). American Society for Engineering Education.

d.School. (n.d.). About. Hasso Plattner Institute of Design at Stanford University. Retrieved from https://dschool.stanford.edu/about

Milovanovic, J., Shealy, T., & Katz, A. (2018). Higher perceived design thinking traits and active learning in design courses motivate engineering students to tackle energy sustainability in their careers. *Sustainability, 13*(22), 12570 (1–14). https://par.nsf.gov/servlets/purl/10312519

WIAL Action Learning. (2018, August 3). *Action learning & design thinking* [Video]. YouTube. https://youtu.be/Vqw5_oSFv6Y

See also: Constructionism, Social Constructivism, Bias Toward Action

3. Advocacy

John Joseph Silvestro

Biases are embedded in everything. Designers of things—from tools to workspaces to institutions—have biases that influence, often unintentionally, the structure and uses of said things. For example, several tools, like cameras, tape-measures, and scissors, were originally designed only for right-handed users. It then falls on individuals to identify these biases and then petition for changes to address them. Advocacy is a process for doing this.

Advocacy is the process of recognizing biases and then promoting changes that help those who have been biased against (Mathieu & George, 2009; Ryder, 2010). Performing advocacy involves recognizing who is disregarded or harmed by an object or community, learning from those affected, and then proposing changes. More precisely, advocacy should follow these steps:

1. **Recognize**: An individual recognizes that a community, space, or institution either fails to acknowledge or harms an individual or group.
2. **Learn**: The individual learns from the group about how they are affected and what they would like to see changed.
3. **Propose**: The individual co-develops changes that will alleviate the challenges the group faces and then presents them to the community.

For the maker movement and individual makers, advocacy is a necessary and productive process. The building of communities and the seeking out and providing of support are central to the maker movement (Hatch, 2014), and both these activities require advocacy to be performed to their fullest potential. To build strong communities, individuals must advocate for new participants as well as new ways of doing things. In order to know when to help and support others, one must monitor the functioning of spaces and communities, identifying the fissures, gaps, and blind spots that cause others unnecessary challenges. Advocacy thus offers a process through which makers can grow and support their own communities and spaces.

For a hypothetical example of advocacy, an individual uses a maker space that is working on expansion plans. The individual recognizes that only twenty-something, single, white males use the current space. So, the individual talks with people who don't use the space, learning accommodations that could be made in the space so they could use it. The individual develops a plan for expanded walkways to accommodate wheelchairs and a play space for children so that working parents can participate. The individual then draws upon his years of using the existing space to present said changes to the group overseeing the expansion.

Performing advocacy also requires understanding and drawing upon one's own position within a community. One must draw upon their standing and credibility to get others to listen and acknowledge the plight of others. Thus, advocacy involves a level of risk: one risks their own position within a community when performing advocacy. Other community members might resist changes and/or challenge the individual advocating for them. Nevertheless, advocacy is essential for developing stronger spaces, communities, and practices, and everyone must be vigilant for the moments and/or spaces in which advocacy is needed.

References

Hatch, M. (2014). *The maker movement manifesto: Rules for innovation in the new world of crafters, hackers, and tinkerers.* McGraw-Hill Education.

Mathieu, P., & George, D. (2009). Not going it alone: Public writing, independent media, and the circulation of homeless advocacy. *College Composition and Communication, 61*(1), 130–149.

Ryder, P. (2010). Public 2.0: Social networking, nonprofits, and the rhetorical work of public making. *Reflections, 10*(1). https://reflectionsjournal.net/wp-content/uploads/CopyrightUpdates/Vol10N1/V10.N1.Ryder_.Phylliss.pdf

Recommended Resources

Cohen, J., Jones, W. M., Smith, S., & Calandra, B. (2017). Makification: Towards a framework for leveraging the maker movement in formal education. *Journal of Educational Multimedia and Hypermedia, 26*(3), 217–229.

Lindeman, N. (2007). Creating knowledge for advocacy: The discourse of research at a conversation organization. *Technical Communication Quarterly, 16*(4), 431–451.

Making a maker space? Guidelines for accessibility and universal design. University of Washington College of Engineering. https://www.washington.edu/doit/making-makerspace-guidelines-accessibility-and-universal-design

Mahoney, K. (2010). Viral advocacy: Networking labor organizing in higher education. *Reflections, 10*(1). https://reflectionsjournal.net/

wp-content/uploads/CopyrightUpdates/Vol10N1/V10.N1.Mahoney. Kevin_.pdf

Nascimento, S., & Pólvora, A. (2018). Maker cultures and the prospects for technological action. *Science and Engineering Ethics, 24*, 927–946. https://doi.org/10.1007/s11948–016–9796–8

Sang, W., & Simpson, A. (2019). The Maker Movement: A global movement for educational change. *International Journal of Science and Mathematics Education, 17*(1), 65–83.

See also: Accessibility, Inclusive Design, Learning Diversity, Ethics

4. Aesthetics

Scott Sundvall

Design is often linked to appearances and appeals. Aesthetics has been commonly understood as the sensory experience, emotions, and judgments related to the visual or physical appearance of objects. When most designers make choices in their design, they consider aesthetics as a criteria of their work as they seek to create things that are pleasing to users. However, aesthetics should not dictate design decisions because it can influence the functionality of a design. Just as the appearance of a chair can affect how people feel when they sit in it, the aesthetics of a website can impact its usability and users' experience.

As a means of appeal, aesthetics can be viewed as a rhetorical device. Since artistic making has been treated under rhetoric, the relationship between design and aesthetics is underlined by the rhetorical tradition. So, this entry provides a brief philosophical tour to frame aesthetics within rhetoric. First, consider three classical definitions of rhetoric:

- "Rhetoric is the art of enchanting the soul" (Plato, 2009);
- "Rhetoric is one great art" (Cicero, 1948); and
- "Rhetoric is the art of speaking well" (Quintilian, 2018).

Additionally, Aristotle (2008) composed an entire treatise entitled *The Art of Rhetoric*. All of these definitive examples gesture to the artistic primacy of rhetoric (rhetoric *qua* art). Four of the five canons of rhetoric—invention, arrangement, style, and delivery—fall under the purview of artistic making.

The aesthetic dimension of rhetoric, wherein there exists an exigency for persuasion, distinguishes it from basic communication. *Suasive force* requires an artistic tact that exceeds the relay of mere information, and it is in this sense that Plato denounces rhetoric (*qua* sophistry) in favor of the philosophical pursuit of "truth" (*aletheia*) (McAdon, 2004). But art (i.e., aesthetics) cannot be reduced to a singular "truth"; as such, nor can rhetoric.

The aesthetic dimension of rhetoric has significantly expanded over the past several decades, including visual rhetoric (Hill & Helmers, 2004; Gries, 2015; Mariani, 2019), multimodal rhetoric (Palmeri, 2012; Murray, 2010; Alexander & Rhodes, 2014), data visualization (Butler, 2011; Beveridge, 2017), and sonic rhetorics (Stone & Ceraso, 2013; Comstock &

Hocks, 2016). Style, as expressed by way of whatever medium or modality, is central to the aesthetic primacy of rhetoric. *How* something is delivered in terms of style and arrangement can matter as much as, if not more than, *what* is rhetorically delivered in content. Consider supporters of Donald Trump and Trumpian rhetoric: they often care less about what he says and more about how he says it—that is, the substance (content) is actually in the style.

Richard Lanham (2007) argues that we have entered into an era of the economic of attention, wherein "fluff" has replaced "stuff." As analogue for rhetorical delivery: it does not matter what is in the box; it matters how the box is wrapped. Victor Vitanza's (1996) desire-aesthetics of the Third Sophistic unfolds rhetoric as the art of endless and multiplicative "and" (*dissoi paralogoi*), rather than the trappings of "either/or" (*dissoi logoi*). As with all art, the aesthetic foundation of rhetoric causes the cup to spilleth over.

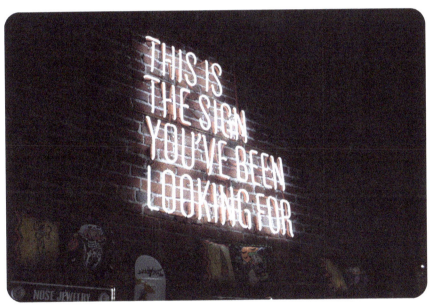

Fig. 4.1. Aesthetics is also concerned with the medium delivering the message. Photo by Austin Chan on Unsplash. https://bit.ly/keyword-fig-4–1

It's important to note that aesthetics is mediated. Gregory Ulmer (2004) contends that we are undergoing an apparatus shift into electracy, a theory that stresses new skills to exploit the communicative potential of new electronic media. Ulmer thinks this phenomenon is similar to the apparatus shift from orality to literacy. In short, alphabetic print is to literacy what digital media is to electracy. But any given apparatus concerns more than the mode or medium of communication; rather, with each apparatus shift there corresponds a shift in metaphysics. Literacy's metaphysics of experiment, method, science, and epistemology are contrasted with electracy's metaphysics of play, style, entertainment, and aesthetics, respectively.

The redoubling of aesthetics in rhetoric in our contemporary moment—as given in Lanham's economics of attention, Vitanza's desire-aesthetics of the Third Sophistic, and Ulmer's electracy—has indeed contributed to our "post-truth" condition. But as rhetoric is an art of persuasion, such only concerns a matter of crafty appropriation. For example, if the *polis* (city-state) rejects the scientific truth of global climate change, then we must find other, more artistic means of persuasion to convince the body politic to save the world (e.g., the Third Sophistic desire-aesthetics of "and" rather than "either/or").

References

Alexander, J., & Rhodes, J. (2014). On multimodality: new media in composition studies. *National Council of Teachers of English.*

Beveridge, A. (2017). Writing through big data: new challenges and possibilities for data-driven arguments. *Composition Forum,* 37.

Butler, S. H. (2011). Teaching rhetoric through data visualization. *Communication Teacher,* 25(3), 131–135.

Cicero. (1948). *On the orator* (E.W. Sutton & H. Rackham, Trans.). Harvard University Press.

Comstock, M., & Hocks, M. E. (2016). The sounds of climate change: sonic rhetoric in the anthropocene, the age of human impact. *Rhetoric Review,* 35(2), 165–175.

Gries, L. (2015). *Still life with rhetoric: a new materialist approach for visual rhetorics.* Utah State University Press.

Hill, C. A., & Helmers, M. H. (2004). *Defining visual rhetorics.* New York: Routledge.

Lanham, R. (2007). *The economics of attention: style and substance in the age of information.* University of Chicago Press.

Mariani, M. (2019). *What images really tell us: visual rhetoric in art, graphic design, and advertising.* Hoaki Books.

McAdon, B. (2004). Plato's denunciation of rhetoric in the Phaedrus. *Rhetoric Review,* 23(1), 21–39.

Murray, J. (2010). *Non-discursive rhetoric: image and affect in multimodal composition.* State University of New York Press.

Palmeri, J. (2012). *Remixing composition: a history of multimodal writing pedagogy.* Southern Illinois University Press.

Plato. (2009). *Phaedrus* (R. Waterfield, Trans). Oxford University Press.

Quintilian. (2018). *Institutes of oratory* (L. Honeycutt, Ed.). Franklin Books.

Stone, J., & Ceraso, S. (2013). Sonic rhetorics: a mashed-up introduction
 in sound. *Harlot*, 9.
Ulmer, G. (2004). *Teletheory*. Atropos Press.
Vitanza, V. (1996). *Negation, subjectivity, and the history of rhetoric*. State University
 of New York Press.

Recommended Resources

Bennett, J., & Zournazi, M. (2020). *Practical aesthetics*. Blooms-
 bury Academic.
Foss, S. K. (2005). Theory of visual rhetoric. In K. Smith, S. Moriarty,
 G. Barbatsis, and K. Kenney (Eds.), *Handbook of visual communication: The-
 ory, methods, and media,* (pp. 141–152). Lawrence Erlbaum.
Kostelnick, C. (2012). Melting-pot ideology, modernist aesthetics, and
 the emergence of graphical conventions: The statistical atlases of the
 United States, 1874–1925. In C.A. Hill and M. Helmer (Eds.), *Defining
 visual rhetorics* (pp. 215–242). Routledge.
Meyer, B., & Verrips, J. (2008). Aesthetics. In D. Morgan (Ed.), *Key words
 in religion, media and culture* (pp. 36–46). Routledge.
Rancière, J. (2010). *Dissensus: On politics and aesthetics*. Bloomsbury Publishing.

See also: Postdigital Aesthetics, Multimodality, Visual Rhetoric, Electracy

5. Ambiguity

Amanda M. May

Within the realm of human experience, it is impossible to perfectly align a constituent and the object to which it refers (Binder et al., 2011). As such, some level of ambiguity, "the absence of a single, clear meaning or the presence of multiple, contradictory meanings" (Ommen, 2016, p. 113), is inevitable. Graphic designs and the designers who create them aim to convey a certain message, yet they grapple with expressing only that message. In other words, ambiguity both renders impossible a unified meaning in graphic design and enables multiple interpretations by designers and audiences alike.

Ambiguity is not unique to visuals. In fact, ambiguity exists in all communicative forms (Ommen, 2016). Further, it is a trait of both verbal and visual arguments (Blair, 2004). Graphic design work typically relies on a combination of both image and verbal communication. Because design is a form of communication (Ommen, 2016), and because it relies on multiple modes like visual and textual, ambiguity in design is inevitable and cannot be completely eliminated. According to one model proposed by Martin J. Eppeler, Jeanne Mengis, and Sabrina Bresciani (2008), ambiguity arises from three causes that function collectively rather than discreetly:

- The visuals themselves, which contain inherent ambiguity arising from the designer's choices.
- The backgrounds of people who participate in the design process or who view a design.
- The interaction between a design, an individual, and the context in which it functions.

Although ambiguity is ubiquitous, a common example exists in figures intentionally designed to be perceived in different ways. Because a design exerts multiple motives—that is, it aims to be perceived in competing ways—viewers can interpret them differently, seeing one object or another (Eppler, Mengis, & Bresciani, 2008). One such figure, the Rubin's Vase ambiguous figure (see Fig. 5.1), is created using black and white: the white shape is a vase, and the black cutouts on either side appear as two faces (Donaldson, 2017).

Fig. 5.1. Image of Rubin's Vase. Source: https://www.illusionsindex.org/i/rubin-s-vase

For makers and designers, ambiguity has both benefits and drawbacks. In discussing screen-based designs, those that are best displayed on a computer or cellular device, Aaris Sherin (2013) favors using less ambiguous graphics that support the branding a designer is attempting to create. Timothy Samara (2014) concurs, noting that the most important aspect of a design is to persuade an audience to the truth in the design's message, and that where decisiveness convinces, ambiguity undermines. The main benefit of ambiguity is its allowance for multiple perspectives. This is true for viewers of a completed design, whose perspectives may emerge over a longer time period (Eppler, Mengis, & Bresciani, 2008) or who may question a design more thoroughly due to ambiguity (Samara, 2014). For those involved in the design process, ambiguity can also yield questions about design (Eppeler, Mengis, & Bresciani, 2008) and result in learning for designers (Canhenha, 2018), but it can also cause conflict between collaborators (Eppeler, Mengis, & Bresciani, 2008).

References

Binder, T., De Michelis, G., Ehn, P., Jacucci, G., Linde, P., & Wagner, I. (2011). *Design things*. MIT Press.

22

Blair, J. A. (2004). The rhetoric of visual arguments. In C.A. Hill and M. Helmers (Eds.), *Defining visual rhetorics* (pp. 40–61). Lawerence Earbaum Associates.

Canhenha, P. (2018, Sept. 23). Ambiguity and the design process [Blog post]. https://uxplanet.org/ambiguity-and-the-design-process-1f2982b6d62a

Donaldson, J. (2017, July). Rubin's vase. In F. Macpherson (Ed.), *The illusions index.* https://www.illusionsindex.org/i/rubin-s-vase

Eppler, M. J., Mengis, J., & Bresciani, S. (2008). Seven types of visual ambiguity: On the merits and risks of multiple interpretations of collaborative visualizations. 2008 12th International Conference Information Visualization, 391–396. https://doi.org/10.1109/IV.2008.47

Ommen, B. (2016). *The politics of the superficial: Visual rhetoric and the protocol of display.* University of Alabama Press.

Samara, T. (2014). *Design elements : Understanding the rules and knowing when to break them* (2nd ed.). Rockport Publishers.

Sherin, A. (2013). *Design elements : Using images to create graphic impact: A graphic style manual for effective image solutions in graphic design.* Rockport Publishers.

Recommended Resources

Barnard, M. (2005). Modernism. In *Graphic design as communication* (pp. 111–136). Routledge.

Christensen, B. T., & Schunn, C. D. (2008). The role and impact of mental simulation in design. *Applied Cognitive Psychology, 22*(1), 1–18.

Leifer, L. L., & Steinert, M. (2011). Dancing with ambiguity: Causality behavior, design thinking, and triple-loop-learning. *Information Knowledge Systems Management, 10*, pp. 151–173. doi:10.3233/IKS-2012-0191

Liu, Y. (1995). Some phenomena of seeing shapes in design. *Design Studies, 16*(2), 367–385.

Taylor, P. (2022, December 7). Designing for ambiguity. https://paulitaylor.com/2022/12/07/16333/

Tseng, W. S.-W. (2018). Can visual ambiguity facilitate design ideation? *International Journal of Technology and Design Education, 28*(2), 523–551. https://doi.org/10.1007/s10798-016-9393-9

See also: Visual Semiotics, Visual Rhetoric, Aesthetics, Postdigital Aesthetics

6. Assemblage

Amanda M. May

Design involves the process of assembling—the drawing together of heterogeneous parts to create a whole. While some consider assemblage as a process (Yancey & McElroy, 2017), other scholars in multiple disciples such as art (Seitz, 1961) and composition (Johnson-Eilola & Selber, 2007) focus on how the term applies to products composed of heterogeneous materials. For the latter, these heterogeneous materials—or other texts—are always explicitly marked. In their case, Johndan Johnson-Eilola and Stuart Selber (2007) used student writing and plagiarism as a backdrop to the discussion about invention, convention, and re-invention.

While it is a helpful starting point, the product/process focus displaces some of the term's theoretical underpinnings, which assemblage theorists like Ian Buchannan (2017) views as detrimental to the term's meaning and the resulting critical applications. As both Buchannan (2017) and Thomas Nail (2017) note, assemblage comes from Gilles Deleuze and Felix Guattari's (2003) theories; their original term, "agencement," has different etymological roots than the term used in many English translations. Nail (2017) distinguishes between these two terms in the following way: "An assemblage is a gathering of things together into unities, an agencement is an arrangement or layout of heterogeneous elements" (p. 22). Readers are encouraged to reference Nail (2017) for the illustrative examples provided to distinguish between agencement and assemblage (as translated), but the crux of the distinction is in the reliance of an essential component in the assembled object—say, compare *an organization* (which has multiple departments and an executive leadership that influences the vitality of the organization, thus resembling the translated assemblage) to *a puzzle* (with no central or essential element; every piece of the puzzle is equally important in forming the full picture, thus resembling the original agencement).

Here, I use assemblage as it is intended by Deleuze and Guattari. Assemblage impacts design in three significant ways. First, as Deleuze and Guattari (2003) note, everything is a middle (*milieu*) in constant flux due to the forces of de/re/territorialization, the emergence, decay, and re-emergence of assemblages as their constituent parts shift as a result of human and non-human forces (p. 11). So, designers and makers need to always consider and accept changing conditions for their projects so they can address the most pressing concerns—hence iterative design.

Second, Deleuze and Guattari's (2003) focus is not on an assemblage's parts but on how those parts—and the assemblage as a whole—function, both presently and in the future. Therefore, designers and makers should pay attention to the holistic experience users may get from interacting with a design rather than a specific feature or function.

Fig. 6.1. Creative groups employ assemblage theory to materialize ideas by creating a "whole" solution through the convergence of "the sum of parts." Photo by Sebastien Bonneval on Unsplash. https://bit.ly/keyword-fig-6–1

Finally, assemblage shows that it is not enough for designers to consider the current context; instead, they must also consider possible futures for their design when it enters the real world. Nail (2017) and Buchanan (2017) both note that Deleuze and Guattari's theory is not intended to ask what an assemblage is, but to ask about how an assemblage is and how it can become; Nail (2017) refers to this as a question of events rather than essences, and Buchanan (2017) highlights an orientation to the future, noting that for Deleuze and Guattari, the question has always been, "given a specific situation, what kind of assemblage would be required to produce it?" (p. 473).

Rather than existing merely to provide solutions, designing with attention to the notion of assemblage allows designers to raise questions. Take another example here: Cartographic maps use visual and textual elements to communicate information about landscapes. Julia Czerniak (1997) differentiates between pictorial maps, those that highlight the natural beauty of landscape, and map-drawings, which render visible aspects of a landscape that may not be apparent; rather than focusing on appearance, map-drawings display how land functions. Designers creating informational maps can use assemblage to consider not only what information is rendered visible by their design choices, but also question how the design choices render these things visible, what they conceal, and what changes would be necessary to emphasize a different issue. As well, assemblage can help designers identify how the map could affect changes in other collectives, such as local communities or other organizations.

References

Buchanan, I. (2017). Assemblage theory, or, the future of an illusion. *Deleuze Studies*, *11*(3), pp. 457–74. doi:10.3366/dls.2017.0276

Czerniak, J. (1997). Challenging the pictorial: Recent landscape practice. *Assemblage*, *34*, pp. 110–120.

Deleuze, G., & Guattari, F. (2003). Rhizome: An introduction. *A thousand plateaus: Capitalism and schizophrenia* (B. Massumi, Trans., pp. 3–25). Continuum. (Original work published 1987.)

Johnson-Eilola, J., & Selber, S. A. (2007). Plagiarism, originality, assemblage. *Computers and Composition*, *24*, pp. 375–403.

Nail, T. (2017). "What is an Assemblage?" *SubStance*, *46*(142): pp. 21–37.

Seitz, W. C. (1961). *The art of assemblage*. New York, NY: The Museum of Modern Art.

Yancey, K. B., & McElroy, S. J. (2017). *Assembling composition*. National Council for the Teachers of English.

Recommended Resources

Edwards, D. W. (2016). *Writing in the flow: Assembling tactical rhetorics in an age of viral circulation*. Doctoral dissertation. Miami University. http://rave.ohiolink.edu/etdc/view?acc_num=miami1465213522

Faris, M. J., Blick, A. M., Labriola, J. T., Hankey, L., May, J., & Mangum, R. T. (2018). Building rhetoric one bit at a time: A case of maker rhetoric with littleBits. *Kairos: A Journal of Rhetoric, Technology, and Pedagogy*, *22*(2). https://kairos.technorhetoric.net/22.2/praxis/faris-et-al/index.html

Johnson-Eilola, J., & Selber, S. A. (2022). Technical communication as assemblage. Technical Communication Quarterly, 32(1), 79–97. https://doi.org/10.1080/10572252.2022.2036815

Lemieux, A., & Rowsell, J. (2020). On the relational autonomy of materials: Entanglements in maker literacies research. *Literacy, 54*(3), 144–152. https://doi.org/10.1111/lit.12226

Tham, J. C. K. (2019). *Multimodality, makerspaces, and the making of a maker pedagogy for technical communication and rhetoric*. Doctoral dissertation. University of Minnesota. https://hdl.handle.net/11299/206361

See also: Invention, Materiality, Curation, Iterative Design

7. Cognitive Dissonance

Johansen Quijano

Design often seeks to achieve consistency. However, mental discomfort can at times be a productive disruption to the design process. The concept of cognitive dissonance finds its origins in the study of psychology. Coined in a 1957 book titled *A Theory of Cognitive Dissonance*, the concept first proposed by Leon Festinger attempts to describe and categorize certain uncomfortable feelings that people experience when either their beliefs or their behaviors are found to be at odds or when new information that challenges an individual's ideas is introduced. These feelings of cognitive dissonance are categorized as a form of psychological stress. Recent research also suggests that cognitive dissonance can also be experienced when deciding between two equally valuable options or beliefs (Chan-Egan, Santos, & Bloom, 2007). The severity of the psychological discomfort experienced is contingent on the individual's personal stake on the belief system being challenged and the ratio of how much of the new information supports previously held paradigms versus the amount of new information that challenges these warrants.

An example of cognitive dissonance in the Maker Movement might happen when one realizes that at times the ethos of the movement runs counter to the way it is practiced. Two of the assumptions of the maker movement are that (1) it is a welcoming and inclusive movement where everyone can learn about DIY culture and that (2) the maker movement is a liberating force that allows everyone to unleash their imagination. However, as Sarah R. Davies (2017) notes in *Hackerspaces: Making the Maker Movement*, this is not always true. Davies notes that even though at first both men and women show interest in maker and hacker spaces for the same reason, it is often the case that they are not treated in the same way. She notes that women who visit maker and hackerspaces are at times treated with condescension, that they are patronized or harassed, or that they are made to feel as if they don't belong. She further notes that maker and hackerspaces are also inaccessible to those without economic resources, writing that despite the belief that these spaces are empowering for everyone, the reality of maker and hackerspaces is that they only empower a specific group of people. When people come to these realizations, Davies notes, they experience some degree of cognitive dissonance. Finally, she notes that those involved in the

Maker Movement might have positive experiences and believe that everyone who visits a maker or hackerspace will have an equally positive experience, but that they might also go through cognitive dissonance upon realizing that these positive experiences are not shared by everyone.

Festinger (1957) suggests that there are four ways one can deal with the uncomfortable sensations caused by cognitive dissonance:

- Individuals can change their practice or belief in light of new information.
- Individuals can justify their behavior by adding conditionals to their belief system.
- Individuals can justify their behavior by changing their beliefs.
- Individuals can ignore the new information.

In the case of an individual realizing that maker and hackerspaces may not be as inclusive and welcoming as they originally believed, then they can:

- Work towards making the space more welcoming.
- Justify the unwelcoming ethos of some spaces by modifying the belief into one along the lines of "it's welcoming to anyone who really wants to work."
- Justify the unwelcoming behavior by accepting it as the status quo: "OK, so it's not welcoming, but that's just the way things are."
- Ignore those who say that the spaces are unwelcome while still believing that they are.

Even though it seems that more often than not individuals opt to justify the practice by adding conditionals to their belief system or by ignoring information (Lawler, 2018), the ideal solution in the hypothetical example outlined above would be to change the practice by working towards making the spaces more welcoming and inclusive by listening to those who feel excluded. Luckily, there is a growing trend of research and practice into making the maker movement inclusive for everyone (Duncan, 2016; Stark-Masters, 2018; Seo, 2019). This should help the movement live up to its theoretical ethos in its practice and reduce cognitive dissonance about the movement throughout the maker community.

References

Davies, S. R. (2017). *Hackerspaces: Making the maker movement*. Polity.

Duncan, S. (2018, December 27). 3 Ways the maker movement can be more inclusive of women and people of color. *EdSurge News*. https://www.edsurge.com/news/2016–05–23-three-ways-the-maker-movement-can-be-more-inclusive-to-women-and-people-of-color

Egan, L. C., Santos, L. R., & Bloom, P. (2007). The origins of cognitive dissonance. *Psychological Science, 18*(11), 978–983.

Festinger, L. (1957). *A theory of cognitive dissonance.* Stanford University Press.

Lawler, M. (2018, February 28). Real-life examples of cognitive dissonance. Everyday Health. https://www.everydayhealth.com/neurology/cognitive-dissonance/real-life-examples-how-we-react/

Seo, J. (2019). Is the maker movement inclusive of anyone?: Three accessibility considerations to invite blind makers to the making world. *TechTrends*, 1–7. doi:10.1007/s11528-019-00377-3

Stark-Masters, A. (2018). How making and maker spaces have contributed to diversity & inclusion in engineering: A [non-traditional] literature review. In *2018 CoNECD – The Collaborative Network for Engineering and Computing Diversity Conference* (Paper ID #23113). American Society for Engineering Education.

Recommended Resources

Bradley, S. (2010, February 23). Cognitive dissonance: How contradictory ideas affect design. https://vanseodesign.com/web-design/cognitive-dissonance/

Kitchener, K. S. (1983). Cognition, metacognition, and epistemic cognition: a three-level model of cognitive processing. *Human Development, 4*, pp. 222–232.

Nikolove, A. (2017, April 1). Design principle: Cognitive dissonance. Medium. https://uxplanet.org/design-principle-cognitive-dissonance-a01dffe81f58

Purzer, S., Hilpert, J. C., & Wertz, R. E. H. (2011). Cognitive dissonance during engineering design. In *Proceedings of 2011 Frontiers in Education Conference (FIE)*, Rapid City, SD, USA (pp. S4F-1-S4F-5). doi: 10.1109/FIE.2011.6142792.

The Decision Lab. (n.d.). Why is it so hard to change someone's beliefs? Cognitive dissonance, explained. https://thedecisionlab.com/biases/cognitive-dissonance

See also: Constructionism, Curiosity, Radical Imagination

8. Constructionism

Merideth Garcia

Current scholarship identifies the origins of the learning theory constructionism in educational making and makerspace style labs. Constructionism posits that knowledge is made, or "constructed," rather than transmitted. In the constructionist paradigm, people make knowledge through interaction with others while creating meaningful artifacts (Kafai, 2006). This formulation of learning was a radical departure from prior theories, which emphasized the transfer of knowledge through direct instruction. Though many of the ideas relevant to constructionism have long been present in philosophies of education, constructionism came to be widely recognized as a theory of learning as a result of Seymour Papert's work with children at the Massachusetts Institute of Technology's (MIT) Artificial Intelligence Laboratory in the 1960s and 1970s (Kafai & Resnick, 2011).

Fig. 8.1. A child creating a 3D rainbow using Play-Doh. Photo by Julietta Watson on Unsplash. https://bit.ly/keyword-fig-8–1

Prior to joining MIT, Papert worked closely with Jean Piaget, a cognitive psychologist who specialized in children's developmental stages of learning. Constructionism draws on Piaget's constructivist theory, which argues that learners construct mental representational models (schemas) of how the world works, integrating new information as they encounter it (Harel & Papert, 1991). While constructivism focuses on the intellectual work that learners do as they try to understand new concepts, constructionism extends this idea to include physical and emotional engagements with the tools and products involved in the learning environment (Papert, 1980). Where constructivism takes curiosity and the desire to learn as default settings, constructionism argues for the importance of a meaningful material object or goal.

In *Mindstorms*, Papert (1980) provides his own childhood fascination with gears as an example, explaining that in addition to demonstrating abstract mathematical concepts, the gear "also connects with the 'body knowledge,' the sensorimotor schemata of a child," and that "there was *feeling, love*, as well as understanding in [his] relationship with gears" (p. viii, emphasis in original). In Papert's view, his emotional and physical engagements with gears at a young age—not easily explained or replicated, but important, nonetheless—were entwined with his lifelong intellectual interest in mathematics and computer science. In both formal and informal educational settings, a constructionist approach involves designing experiences that allow learners to discover, create, and share artifacts that are personally meaningful. This emphasis on learning through community and creativity is directly relevant to maker culture, which designs spaces to promote processes of building, participating, learning, and sharing (Hatch, 2013).

Constructionism also functions as a research methodology, focusing on the layers of meaning that impact the design and reception of a product or idea. Critics of constructionism as a methodology argue that some forms of it seem to conflict with scientific realities. How does one identify facts, if knowledge is something that humans *make* and not something that simply *is*? While constructionist research does not claim that empirical realities are only what people think they are, it does focus on how people encounter realities and make narratives that account for them. As a result, it is less concerned with producing generalizable results than with investigating local or personal understandings and how they came to be (Neimeyer & Torres, 2001).

References

Harel, I., & Papert, S. (Eds.). (1991). *Constructionism*. Ablex Publishing Corporation.

Hatch, M. (2014). *The maker movement manifesto: rules for innovation in the new world of crafters, hackers, and tinkerers*. McGraw-Hill Education.

Kafai, Y. B. (2006). Constructionism. In K. Sawyer (Ed.), *Cambridge handbook of the learning sciences* (pp. 35–46). Cambridge University Press.

Kafai, Y. B., & Resnick, M. (2011). *Constructionism in practice: Designing, thinking, and learning in a digital world*. Routledge.

Neimeyer, R. A., & Torres, C. (2014). Constructivism and constructionism: Methodology. In J. D. Wright (Ed.), *International encyclopedia of social and behavioral sciences* (2nd Ed., Vol 4, pp. 724–728). Elsevier.

Papert, S. (1980). *Mindstorms: Children, computers, and powerful ideas*. Basic Books.

Recommended Resources

Becker, S. (2016). Developing pedagogy for the creation of a school makerspace: Building on constructionism, design thinking, and the Reggio Emilia approach. *The Journal of Educational Thought, 49*(2), 192–209. https://www.jstor.org/stable/26372370

Hartle, R. T., Baviskar, S., & Smith, R. (2012). A field guide to constructivism in the college science classroom: Four essential criteria and a guide to their usage. *Bioscene, 38*, 31–34.

Holbert, N., Berland, M., & Kafai, Y. B. (Eds.). (2020). *Designing constructionist futures: The art, theory, and practice of learning designs*. MIT Press.

Kelter, J., Peel, A., Bain, C., Anton, G., Dabholkar, S., Horn, M. S., & Wilensky, U. (2021). Constructionist co-design: A dual approach to curriculum and professional development. *British Journal of Educational Technology, 52*(3), 1043–1059. https://doi.org/10.1111/bjet.13084

Thanapornsangsuth, S., & Holbert, N. (2020). Culturally relevant constructionist design: exploring the role of community in identity development. *Information and Learning, 121*(11/12), 847–867. DOI 10.1108/ILS-02-2020-0024

See also: Social Constructivism, Active Learning, Play/Playful

9. Context and Contextual Design

Thomas M. Geary

In making and design thinking, as with all forms of communication, the content or product is always situated within a larger context, the frame of historical and current elements that vary in each situation and for each user. The circumstances that shape a context in any given situation may include surrounding events, institutional norms, community, space, and time. Sara L. Beckman and Michael Barry (2007) describe context as occurring on multiple levels: "immediate physical and situational surroundings, language, character, culture, and history all provide a basis for the meaning and significance attached to roles and behavior" (p. 31). Thus, the complex reality of the twenty-first century workplace has resulted in design approaches that foreground context in an effort to adapt products and services to user needs.

Fig. 9.1. Technology design must consider the context in which technological interactions take place. Photo by Valerie V on Unsplash. https://bit.ly/keyword-fig-9–1

Developed by Hugh Beyer and Karen Holtzblatt in 1988, contextual design is a process of designing by situating in the everyday life of the target users. Contextual design has become the industry standard in various fields and widely adopted in education, urban design, management, healthcare, systems analysis, mobile application development, and human-computer interaction. The collaborative approach goes beyond the traditional process of idea generation, testing, and analysis to determine user experience, cultural fit, and appropriateness for each environment and utilize those findings in development. David Dunne and Roger Martin (2006) stress the importance of empathy with others to understand their needs and experiences. Echoing Don Norman's (2002) sentiments, they write, "Because we tend to project our own rationalizations and beliefs onto others, designers can become isolated from users' needs and interests, and functionality can suffer" (p. 519). Awareness and attention to context and situatedness results in a designerly approach and fosters successful collaboration.

To understand one's "core motives" and context—their stories, successes, failures, aspirations, and concerns—Holtzblatt and Beyer (2017) utilize contextual inquiry, a field research process of close observation (p. 50). The probing, analytical contextual inquiry assists designers in understanding how a culture operates, how products are used, and how decisions are made (Beckman & Berry, 2007). Designers attempt to identify barriers within the environment or product usage that may not be apparent to users. A proper understanding and analysis of the contextual complexities can result in discoveries and creative approaches. Richard Buchanan (1992) contends that without attention to context, "Ideas are then forced onto a situation rather than discovered in the particularities and possibilities of that situation" (p. 13).

In a TED talk with Charlie Rose, Larry Page (2014) discusses Google's contextual approach toward product design, from increased accessibility and usability to adaptation of software and hardware that understands users. Improving a computer's ability to understand a user's context and therefore needs in any given situation is the next evolution for the company, but, as Page acknowledges, "It's still very, very clunky" due to the complex nature of context. Contextual design is an ongoing area for inquiry in making and design studies.

References

Beckman, S. L., & Barry, M. (2007). Innovation as a learning process: Embedded design thinking. *California Review Management, 50(1)*, 25–56.

Beyer, H., & Holtzblatt, K. (1998). *Contextual design: Defining customer-centered systems*. Morgan Kaufmann.

Buchanan, R. (1992). Wicked problems in design thinking. *Design Issues, 8*(2), 5–21.

Dunne, D., & Martin, R. (2006). Design thinking and how it will change management education: An interview and discussion. *Academy of Management Learning & Education, 5*(4), 512–523.

Holtzblatt, K., & Beyer, H. (2017). *Contextual design: Design for life.* (2nd edition). Morgan Kaufmann.

Norman, D. A. (2002). *The design of everyday things.* Basic Books.

Page, L. (2014). Where's Google going next? TED2014. https://www.ted.com/talks/larry_page_where_s_google_going_next/

Recommended Resources

Aranda-Jan, C. B., Japtap, S., & Moultrie, J. (2016). Towards a framework for holistic contextual design for low-resource settings. *International Journal of Design, 10*(3), 43–63.

Augstein, M., Neumayr, T., Pimminger, S., Ebner, C., Altmann, J., & Kurschl, W. (2018). Contextual design in industrial settings: Experiences and recommendations. In *Proceedings of the 20th International Conference on Enterprise Information Systems (ICEIS 2018) - Volume 2* (pp. 429–440).

Löfflerr, D., Wallmann-Sperlich, B., Wan, J., Knött, J., Vogel, A., & Hurtienne, J. (2015). Office ergonomics driven by contextual design. *Ergonomics in Design, 23*(3), 31–35. DOI: 10.1177/1064804615585409

Ostuzzi, F., De Couvreur, L., Detand, j., & Saldien, J. (2017). From design for one to open-ended design: Experiments on understanding how to open up contextual design solutions. *The Design Journal, 20*(1), S3873–S3883. https://doi.org/10.1080/14606925.2017.1352890

Rosenfeld Halverson, E., & Sheridan, K. (2014). The maker movement in education. *Harvard Educational Review, 84*(4), 495–504.

See also: Local, Invention, User Requirements

10. Creative Confidence

Kamila Albert

In 1973, Bernie Roth, a founder of the Hasso Plattner Institute of Design at Stanford, delivered a series of lectures on the design process and creativity. Early on, the lectures dispelled myths about creativity as a talent associated with some form of genius. Instead, Roth wrote, everyone "possess[es] some ability to think creatively and to engage themselves in imaginative and innovative efforts." This idea has developed into the concept of creative confidence. Creative confidence may be defined as a person's belief in their ability to generate ideas, solutions, or approaches (Kelley & Kelley, 2013) and has become a central aspect of design thinking. With roots in psychology, creative confidence is often compared to Albert Bandura's (1977) concept of self-efficacy and connected to Carol Dweck's (2006) work on growth vs. fixed mindsets.

David Kelley and Tom Kelley (2013) compare creative confidence to a muscle that can be "strengthened and nurtured through effort and experience." Creative confidence is often related to Bandura's (1977) concept of self-efficacy or the notion that "our belief systems affect our actions, goals, and perception." For those with lower levels of self-efficacy, Bandura believed in a process of guided mastery, or a series of small successes, to help people gain courage. Likewise, people who doubt their creative ability can change their views through guided principles and exercises. Creative agency or the ability to apply one's creativity in real-world contexts has been presented as a complement to self-efficacy (Royalty & Roth, 2016). Creative confidence has also been connected to Dweck's (2006) research on growth and fixed mindsets. Developing creative confidence is based largely on a growth mindset or the belief that a person's innovation skills and creative abilities are not fixed. People with a growth mindset imagine that their abilities can be expanded through effort and experience. Contrarily, people with a fixed mindset believe that a person's qualities are unchangeable (Dweck, 2006).

Creative growth mindset also connects to how people respond to failure, which is a common occurrence in innovative practices (Royalty & Roth, 2016). Failure is "baked into the creative process" and the way that a person responds to failure plays a role in their sense of creative confidence (Kelley & Kelley, 2013). People with growth mindsets respond better to failure because they see those attempts as ways to learn what did not work. Design thinking, a human-centered approach to design and innovation, attempts to foster creative confidence through doing and re-orienting ideas about creativity and failure. In design thinking, failures are presented as learning opportunities because the relationship between creative application and failure is inevitable. For an example, see "From Fear to Courage" (https://www.creativeconfidence.com/chapters/chapter-2).

Creative confidence has become a central concept of design thinking. More so, it suggests that a person's creative capacity can be cultivated and that anyone can "generate new ideas, solutions, or approaches" (Kelley & Kelley, 2013). Individuals who believe in their creative confidence or ability to effect change "are more likely to accomplish what they set out to do in the world" (Kelley & Kelley, 2013).

References

Bandura, A. (1977). Self-efficacy: Toward a unifying theory of behavioral change. *Psychological Review, 84*(2), 191–215.

Dweck, C. (2006). *Mindset: the new psychology of success.* Random House.

Kelley, D., & Kelley, T. (2013) *Creative confidence: Unleashing the creative potential within us all.* Random House.

Royalty, A., & Roth, B. (2016). Developing design thinking metrics as a driver of creative innovation. In H. Plattner, C. Meinel, and L. Leifer (Eds.), *Design thinking research: Making design thinking foundational* (pp. 171–186). Springer Press.

Recommended Resources

Karwowski, M., Han, M.-H., & Beghetto, R. A. (2019). Toward dynamizing the measurement of creative confidence beliefs. *Psychology of Aesthetics, Creativity, and the Arts, 13*(2), 193–202. https://doi.org/10.1037/aca0000229

Plattner, H., Meinel, C., & Leifer, L. (2016). *Design thinking research: Making design thinking foundational.* Springer Press.

Roth, B. (1973). *Design process and creativity.* Stanford Institute of Design Online Resources. https://static1.squarespace.com/static/57c6b79629687fde090a0fdd/t/590133396a4963a462c-680cd/1493250903831/Design+Process+and+Creativity+B+Roth+Small.pdf

Rauth, I., Köppen, E., Jobst, B., & Meinel, C. (2010). Design thinking: An educational model towards creative confidence. In *Proceedings of the 1st International Conference on Design Creativity, ICDC 2010* (pp. 1–8). https://

www.designsociety.org/download-publication/30267/Design+Think-ing%3A+An+Educational+Model+towards+Creative+Confidence

Sadley, J., Shluzas, L., Blikstein P., & Katila R. (2016). Building blocks of the maker movement: Modularity enhances creative confidence during prototyping. In H. Plattner, C. Meinel, and L. Leifer (Eds.), *Design thinking research: Making design thinking foundational* (pp. 141–156). Springer Press.

See also: Play/Playful, Curiosity, Creative Confidence, Cognitive Dissonance

11. Cultural Intelligence

Nitya Pandey

In simple words, culture is the collective practice of shared values passed along from one generation to another. In today's world of globalization, it is common for people of different cultures to meet virtually or in person at various institutions and spaces. Therefore, in order to be able to create a cordial and professional atmosphere, it is important that these people are sensitive to the multiple dimensions of different cultures. It is necessary for them to understand that each culture has some core values that are important to people who feel a sense of belonging to that specific cultural group.

Cultural intelligence, in that sense, is the ability to make sense of the foreign and the unfamiliar, and cultivate the thought and attitude that fairly addresses exclusion, awkwardness, and hostility. This trait is extremely crucial because it helps people react appropriately to certain situations and make the most out of them. When you are aware of how a culture operates, you gain the ability to form strategies that help you deal with certain circumstances in relation to that culture. But even when you happen to find yourself in a situation that is completely novel to you, if you have cultural intelligence, you will succeed in navigating your way through it.

Here is an example of cultural intelligence in marketing, about a candy, its wrapper, and its consumers. Many years ago, before cellphones and social media, a foreign candy became very popular in Kathmandu, Nepal. It was a white square block, mushy in texture, and extremely sweet in taste. It was loved by kids as well as adults. It had a picture of a cow on its wrapper and something written in a script that was unfamiliar to most people in town.

Kathmandu is largely a Hindu/Buddhist community and in Hinduism, beef is strictly prohibited. So, when someone, at some point, spread the rumor that the candy was made from cow fat, and even had a picture of a cow on its wrapper, people were shocked and disgusted. The news instantly created ripples in the candy market. There were some people who tried to read the foreign script and make sense of what was written on it. But it was too late by then because the candy had acquired a bad reputation. Soon, everybody stopped buying the candy and eventually, it phased out from the market.

This is an example where a product that had a great deal of potential failed miserably due to the lack of cultural intelligence on both ends of the marketing spectrum. If the candy makers had analyzed their target audience and presented the information in both languages, or a language understood more widely in the area, the misunderstandings could have been avoided. Similarly, if the consumers had displayed a little more patience, and attempted to investigate the truth before completely rejecting the candy, perhaps they would have realized that the face of the cow on the wrapper only meant that this was a milk candy.

Fig. 11.1. Animal and human relations as exemplified in the Nepali example. Photo by Monthaye on Unsplash. https://bit.ly/keyword-fig-11-1

Recommended Resources

Ang, S., & Van Dyne, L. (2008). *Handbook of cultural intelligence: Theory measurement and application*. Routledge.

Cultural Intelligence Center. (n.d.). About cultural intelligence. https://culturalq.com/about-cultural-intelligence/

Earley, P. C. (2003). *Cultural intelligence: Individual interactions across cultures*. Stanford University Press.

Earley, P. C., & Mosakowski, E. (2004) Cultural intelligence. *Harvard Business Review*. https://hbr.org/2004/10/cultural-intelligence

Ott, D. L., & Michailova, S. (2016). Cultural intelligence: A review and new research avenues. *International Journal of Management Reviews,* 20(1), 99–199.

See also: Context & Contextual Design, Brand/Branding, Local

12. Curiosity

Alexandra Catá-Ross

"As a trained, tenured scholar in the humanities, let me tell you, "What the fuck?" is the underlying research question to much excellent scholarship" —Dr. Carly Kocurek

Curiosity is an interesting word in this collection. It's not a buzzword like innovation, concretely defined like project management, or even central to a field like user experience. Curiosity is an outlier in all of these aspects, and, yet, is so critical to our motivations and interests as designers, tinkerers, and communicators.

Curiosity has been researched empirically throughout the decades and generated several definitions and types. Sophie von Strumm et al.'s (2011) study on the role of intellect in student and academic performance discusses epistemic curiosity as the "desire or hunger for human knowledge" (p. 577). This, combined with cognition and the Typical Intellectual Engagement scale, creates "trait constructs that describe tendencies to seek out, engage in, enjoy, and pursue opportunities for effortful cognitive activity—in short, intellectual curiosity" (p. 578).

Jordan Litman and Charles D. Spielberger's (2003) definition broadly defines curiosity as the desire to gain new knowledge and sensory experiences that motivate "exploratory behavior" (Reio et al., 2006, p. 118). Exploratory behavior, in turn, generates "cognitive, social, emotional, spiritual, and physical development" (p. 117). Reio et al. break down curiosity into two types: information seeking and sensory seeking. Cognitive curiosity drives information-seeking exploratory behavior, while sensory curiosity drives sensation-seeking exploratory behavior (p. 118). These types suit different purposes and needs: "Information-seeking curiosity might help answer questions about affordances related to specific objects or events, whereas sensory curiosity might motivate spontaneous seeking of new opportunities for action" (p. 118).

The idea of intellectual, information-seeking, sensory-seeking, and epistemic curiosity certainly plays a role in academic pursuits and research decisions, particularly in relation to critical making and technical communication. Alyssa Arbuckle and Alex Christie (2015) describe critical making as not just thinking about production and "[closing] the loop between material production and theoretical engagement," but that by engaging with materiality in this way, scholars are also producing social knowledge. "This is especially evident when research outputs, including data sets, visualizations, and 3-D printable objects, are available for uptake and reuse by scholars in multiple fields" (p. 10). Not only does critical making allow for new forms of research, but the ways in which research is generated, methodologically implemented, and published are all varied and different from traditional publishing outlets. This is valuable for technical communication as scholars and practitioners are increasingly becoming more hands-on with various types of digital media, technologies, and materials. Even in my own experiences as a technical communication practitioner and academic, curiosity is the basis for thinking critically about how a technology functions and how it affects users not just at a practical level, but also at a social, cognitive, and theoretical level as well.

Curiosity may seem like an out-of-place word on this list; however, it drives our academic and personal motivations, desires, and goals. In any project that allows researchers to be hands-on with their study group, the tactile and material interactions provoke intellectual and sensory understandings of the way things are constructed, used, distributed, and published, as well as demonstrating their impacts on the world and people around them. At the same time, epistemic curiosity is what makes us scratch our heads, puzzle over ideas, and see pieces of information in a different light.

References

Arbuckle, A., & Christie, A. (2015). Intersections between social knowledge creation and critical making. *Scholarly and Research Communication*, 6(3). http://dx.doi.org/10.22230/src.2015v6n3a200

Litman, J., & Spielberger, C. D. (2003). Measuring epistemic curiosity and its diversive and specific components. *Journal of Personality Assessment*, *80*(1), 75–86.

Reio, T. G., Jr., Petrosko, J. M., Wiswell, A. K., & Thongsukmag, J. (2006). The measurement and conceptualization of curiosity. *The Journal of Genetic Psychology*, *167*(2), 117–35. https://doi.org/10.3200/GNTP.167.2.117–135

sparklebliss (2019, June 15). As a trained, tenured scholar in the humanities, let me tell you, "What the fuck?" is the underlying research question to much excellent scholarship. [Tweet]. https://twitter.com/sparklebliss/status/1139146172113870848

von Stumm, S., Hell, B., & Chamorro-Premuzic, T. (2011). The hungry mind: Intellectual curiosity is the third pillar of academic per-

formance. *Perspectives on Psychological Science, 6*(6), 574–588. https://doi.
org/10.1177/1745691611421204

Recommended Resources

Gross, M. E., Zedelius, C. M., & Schooler, J. W. (2020). Cultivating an understanding of curiosity as a seed for creativity. *Current Opinion in Behavioral Sciences, 35*, 77–82. https://doi.org/10.1016/j.cobeha.2020.07.015

Kidd, C., & Hayden B. Y. (2015). The psychology and neuroscience of curiosity. *Neuron, 88*(3), 449–460. https://doi.org/10.1016/j.neuron.2015.09.010

Lopez Cervera, R., Wang, M. Z., & Hayden, B. Y. (2020). Systems neuroscience of curiosity. *Current Opinion in Behavioral Sciences, 35*, 48–55. https://doi.org/10.1016/j.cobeha.2020.06.011

Shankar, A., & Zurn, P. (Eds.). (2020). *Curiosity studies: A new ecology of knowledge.* University of Minnesota Press.

Wagstaff, M. F., Flores, G. L., Ahmed, R., & Villanueva, S. (2020). Measures of curiosity: A literature review. *Human Resources Development Quarterly, 32*(3), 363–389. https://doi.org/10.1002/hrdq.21417

See also: Creative Confidence, Cognitive Dissonance, Play/Playful, Critical Making

13. Digital Rhetoric

Kevin Brock

Digital rhetoric is a term used to describe the study and practice of communication and meaning-making occurring through and relating to electronic, and especially digital, media (Losh, 2009; Eyman, 2015). This includes both conventional communicative forms and genres that have been adapted to digital platforms as well as new and emerging forms and genres. Examples span from blogs, discussion forums, and digitized broadcasts of audio or video to image-heavy memes, software code, and practices regarding technological literacy and its dissemination.

While digital rhetoric is not limited to online or other networked media, such venues provide easily identifiable and frequently accessible examples and kinds of, as well as opportunities for, relevant communication. In particular, the most popular social media platforms (e.g., Facebook, Instagram, and Twitter) serve as spaces for highly visible and wide-reaching digital rhetoric practices.

Fig. 13.1. Design of interface is highly rhetorical and can influence users' behaviors and decision making. Photo by freestocks on Unsplash. https://bit.ly/keyword-fig-13–1

As an area of study, digital rhetoric encompasses the examination of how individuals and communities compose meaningfully in and about digital technologies and disseminate their messages to and with various audiences. This area overlaps, in some cases significantly, with work in related disciplines and areas of interest, from communication to composition to the digital humanities to sociology.

Some scholars' research focuses primarily on retrospective or contemporary analysis of particular practices (such as in civic, academic, or professional spheres) and how they impact various scales of community, culture, and society (Warnick, 2007; Hess & Davisson, 2018). Some study the rhetorical dimensions of and considerations regarding the development and use of platforms and tools for digital communication, scrutinizing the influences and biases that lead to the construction and application thereof (Banks, 2011; Vee, 2017; Noble, 2018). Some focus on the instruction of such tools, genres, and rhetorical principles so that students—or writers in general—might more effectively employ them for desired ends (Lanham, 1993; Ridolfo & Davidson, 2015; Rieder, 2017).

Expertise with, and rhetorical awareness regarding, a variety of digital tools—expressly discursive and otherwise—contributed significantly to the success of Barack Obama's re-election campaign in 2012. During that campaign, his team of tech-savvy volunteers developed an open-source platform called "Narwhal" that enabled them to share important voter information with volunteers so as to more effectively identify and target voter interests as well as facilitate coordination of local groups' efforts to reach potential voters, both in-person and through social media channels (Madrigal, 2012). Conversely, opponent Mitt Romney's campaign staff, with the aid of Microsoft and an undisclosed consulting firm, developed a much more limited and proprietary-based platform (named "ORCA," after the natural predator of the narwhal) that suffered from a number of system crashes, confusing or otherwise non-intuitive interfaces and tools, and poor team training of its use, all of which led to the platform's inability to distribute important information to volunteers and voters in a timely or helpful manner (Gallagher, 2012).

References

Gallagher, S. (2012). Inside Team Romney's whale of an IT meltdown. *Ars Technica.* https://arstechnica.com/information-technology/2012/11/inside-team-romneys-whale-of-an-it-meltdown/

Madrigal, A. C. (2012). When the nerds go marching in. *The Atlantic.* https://www.theatlantic.com/technology/archive/2012/11/when-the-nerds-go-marching-in/265325/

Recommended Resources

Banks, A. J. (2011). *Digital griots: African American rhetoric in a multimedia age.* Southern Illinois University Press.

Brooke, C. G. (2009). *Lingua fracta: Towards a rhetoric of new media.* Hampton Press.

Eyman, D. (2015). *Digital rhetoric: Theory, method, practice.* University of Michigan Press. doi:10.3998/dh.13030181.0001.001.

Hess, A., & Davisson, A. (Eds.). (2017). *Theorizing digital rhetoric.* Routledge.

Lanham, R. (1993). *The electronic word: Democracy, technology, and the arts.* University of Chicago Press.

Losh, E. (2009). *Virtualpolitik: An electronic history of government media–making in a time of war, scandal, disaster, miscommunication, and mistakes.* MIT Press.

Noble, S. (2018). *Algorithms of oppression: How search engines reinforce racism.* New York University Press.

Ridolfo, J., & Hart-Davidson, W. (Eds.). (2015). *Rhetoric and the digital humanities.* University of Chicago Press.

Rieder, D. M. (2017). *Suasive iterations: Rhetoric, writing, & physical computing.* Parlor Press.

Vee, A. (2017). *Coding literacy: How computer programming is changing writing.* MIT Press.

Warnick, B. (2007). *Rhetoric online: Persuasion and politics on the World Wide Web.* Peter Lang.

See also: Postdigital Aesthetics, Coding, Electracy, Visual Rhetoric

14. Electracy

Kristopher Purzycki

Although digital tools have been incorporated into educational curricula for at least two generations of students, instruction towards the techniques, methods, and impacts of these tools has been slower to develop. Scholars paying attention to the emergence of digital media—including those teaching or practicing design—quickly recognized that these new forms of communication would require new fluencies. In *Internet Invention*, Greg Ulmer (2003) argued that "electracy" would be to electronic media as literacy was to print media. In the nearly two decades since, neologism has become an explicit metric for praxis and curriculum design even while becoming a more understated tentpole for the digital humanities.

Fig. 14.1. A young user of computing technology. Photo by Thomas Park on Unsplash. https://bit.ly/keyword-fig-14–1

According to Ulmer, electracy offers a perspective towards media that situates the emergent apparatus as a social machine (2003, p. 28). In contrast to Marshall McLuhan's (1964) more deterministic interpretation of the media influenced society, electracy contends that all forms of technology are relational to the communities from which they emerge. The practice of electracy is akin to entertainment and the creation of media forms (Holmevik, 2012, p. 7). In light of this, electracy has been picked up primarily by those embracing game-based pedagogies. Jan Rune Holmevik (2012), for example, argues that to be electrate is a form of practice that bridges pleasure and pain as a way to "invent electracy at the intersection of play and the human question" (p. 17).

The heuristic of electracy continues to assert pedagogical value, particularly in rhetoric and writing studies where design-oriented curricula have been adopted. Sarah Arroyo's (2013) *Participatory Composition*, for instance, expands on electrate practices in video culture. Participatory culture, as embodied in electracy and writing, requires students' *intentional* engagement with technologies not only as objective tools but as platforms that impact and resonate through audiences (p. 121). Other educators like Ola Erstad (2003) have argued that electracy is more than simple participation in cultural forms but a means for (guided) student empowerment.

Despite being a pedagogical heuristic that may seem quaint at first glance, electracy implicates how complicated and ingrained our relationships with technology have become. In design-oriented classrooms, to be electrate is to also comprehend the processes by which our technologies (as well as their users) participate in culture and how articulating those processes can reveal their affect and impact on our communities. Through electracy, we make elaborate the obfuscated and unanticipated connections—the social strata—of a society immersed in the spectacle.

References

Arroyo, S. (2013). *Participatory composition: Video culture, writing, and electracy.* Southern Illinois University Press.

Erstad, O. (2003). Electracy as empowerment: Student activities in learning environments using technology. *Young, 11*(1), 11–28.

Holmevik, J. (2012). *Inter/vention: Free play in the age of electracy.* MIT Press.

McLuhan, M. (1964). *Understanding media: The extensions of man.* Signet Books.

Ulmer, G. (2003). *Internet invention: From literacy to electracy.* Longman Publishers.

Recommended Resources

Boyle, C., Brown, Jr., J. J., & Ceraso, S. (2018). The digital: Rhetoric behind and beyond the screen. *Rhetoric Society Quarterly, 48*(3), 251–259. https://doi.org/10.1080/02773945.2018.1454187

Morey, S. (2015). *Rhetorical delivery and digital technologies: Networks, affect, electracy.* Routledge.

Porter, J. E. (2009). Recovering delivery for digital rhetoric. *Computers and Composition, 26*(4), 207–224. https://doi.org/10.1016/j.compcom.2009.09.004

Rice, J., & O'Gorman, M. (2008). *New media/new methods: The academic turn from literacy to electracy (New media theory)*. Parlor Press.

Zappen, J. P. (2005). Digital rhetoric: Toward an integrated theory. *Technical Communication Quarterly, 14*(3), 319–325. https://doi.org/10.1207/s15427625tcq1403_10

See also: Digital Rhetoric, Visual Rhetoric, Coding, Postdigital Aesthetics

15. Embodiment

Heather Listhartke

B oth acts of designing and making are embodied experiences. Various definitions exist centered around the idea of using one's own cultural context and awareness of their own body and acts of cognition to produce knowledge or a product. In rhetorical studies, scholars have explicated embodiment as "tectonic reverberations" (Royster & Kirsch, 2012) that is influencing the way we see and *be* in the world—through lived experiences (Banks, 2003), physical inhabitation of places and spaces, sexuality (Butler, 1993), disability (Dolmage, 2014; Price, 2015), and other intersectional experiences (e.g., Crenshaw, 1989; Gay, 2017; Hayles, 1999; Schalk, 2018). A. Abby Knoblauch and Marie E. Moeller (2022) considered the body as a text that interfaces with the world in rhetorical representations and multifaceted engagement. As now a given part of pedagogical instruction and underlying theory of the process of composing, especially in new media or multimodal composing, embodiment is an important theoretical framework for understanding how we make things as embodied beings in the world.

Fig. 15.1. Making and composing involve the body and intersectional experiences of the maker. Photo by Jen Theodore on Unsplash. https://bit.ly/keyword-fig-15–1

Kristen Arola and Anne Frances Wysocki (2012) in their edited book, *Composing(Media) = Composing(Embodiment)*, point to this history of use in making the connection between embodiment and practice composing media. They argue, "We come to be always already embedded—embodied—in mediation. Our relations with our media matter, in other words, and (this is one lesson we take from the philosophers and thinkers just mentioned) we therefore need to consider our engagements with our media if we and the people in our classes are to learn about our embodiment and so what we consider ourselves to be and to be able to do in our worlds" (Arola & Wysocki, 2012).

Though Arola and Wysocki's book is one of the few in rhetoric and composition that focuses on embodiment in composition, their work takes from a handful of key texts before them. As they point out earlier in their book, the term was first termed in 1945 within the work of Maurice Merleau-Ponty, defining it as "the experience of one's body as a unified potentiality or capacity for acting in the world through time and space" (Arola & Wysocki, 2012).

Within writing studies, one of the first to emphasize the importance of embodiment is Christina Haas (1996) through her book *Writing Technology: Studies on the Materiality of Literacy*. She cites the works of Jay Bolter (1993) and Eric Havelock (1988), in an effort to bring together the ideas of embodiment and materiality of technology as a foundation and framework for the results and arguments that she presents on the student practices of digital literacy. Specifically, she says that cultural tools and cognitive activity construct one another through the embodied actions of human beings. She goes further to argue for an embodied, everyday practice in approaching the relationships between technologies and literacy (Haas, 1996).

We can see the echoes of this definition in Arola and Wysocki's definition where they say, "'Embodiment' in this understanding, calls us to attend to what we just simply do, day to day, moving about, communicating with others, using objects that we simply use in order to make things happen" (Arola & Wysocki, 2012).

We can see this practice in the popular pedagogical instructions that focus on metacognition and students' use of felt sense, first defined in the article by Sondra Perl (1980) entitled "Understanding Composing." Perl describes this felt sense as attention not to the words on the page or the topic "but to feelings or non-verbalized perceptions that surround the words" (1980, p. 365). Felt sense, in other words, brings the writer's physical and emotional feelings, mental cognition, and personal (cultural) experiences together to drive their writing style.

Recent scholars in the field have connected more attention to cultural contexts and using them as a necessary part of the composition style and voice we ask for from students, such as Janine Butler's (2016) demonstration of embodied practices in composition pedagogy and composing processes through the examples of ASL music videos. She specifically states, "by rhetorically analyzing how ASL music videos synchronize rhetorical meaning across modes, students can begin to re-conceptualize the affordances of modes and better appreciate how different modes can coordinate in reaching the senses" (Butler, 2016) Essentially, if students use their own embodied experiences to compose texts, they can more effectively create texts that get their audience to have embodied experiences while reading. Similarly, designers and makers can examine their own embodiment in the creative process to better understand the interconnection of minds, bodies, media, and meanings.

References

Arola, K. L., & Wysocki, A. F. (2012). *Composing(media) = composing(embodiment): Bodies, technologies, writing, the teaching of writing*. Utah State University Press.

Banks, W. P. (2003). Writing through the body: Disruptions and "personal" writing. *College English, 66*(1), 21–40.

Bolter, J. D. (1993). Alone and together in the electronic bazaar. *Computers and Composition, 10*(2), 5–17.

Butler, J. (1993). *Bodies that matter: On the discursive limits of "sex."* Routledge.

Butler, J. (2016). Where access meets multimodality: The case of ASL music videos. *Kairos, 21*(1). http://kairos.technorhetoric.net/21.1/

Crenshaw, K. (1989). Demarginalizing the intersection of race and sex: A black feminist critique of antidiscrimination doctrine, feminist theory, and antiracist politics. *University of Chicago Legal Forum, 1989*(8), 139–167.

Dolmage, J. (2014). *Disability rhetoric*. Syracuse University Press.

Gay, R. (2017). *Hunger: A memoir of (my) body*. Corsair.

Haas, C. (2009). *Writing technology studies on the materiality of literacy*. Routledge.

Havelock, E. (1988). *The muse learns to write: Reflections on orality and literacy from antiquity to the present*. Yale University Press.

Hayles, N. K. (1999). *How we became posthuman: Virtual bodies in cybernetics, literature, and informatics*. University of Chicago Press.

Johnson, M. (2017). *Embodied mind, meaning, and reason: How our bodies give rise to understanding*. University of Chicago Press.

Knoblauch, A. A., & Moeller, M. E. (Eds.). (2022). *Bodies of knowledge: Embodied rhetorics in theory and practice*. Utah State University Press.

Perl, S. (1980). Understanding composing. *College Composition and Communication, 31*(4), 363–369. doi:10.2307/356586

Price, M. (2015). The bodymind problem and the possibilities of pain. *Hypatia, 30*(1), 270–284.

Roysters, J. J., & Kirsch, G. E. (2012). *Feminist rhetorical practices: New horizons for rhetoric, composition, and literacy studies.* Southern Illinois University Press.

Schalk, S. (2018). *Bodyminds reimagined: (Dis)Ability, race, and gender in Black women's speculative fiction.* Duke University Press.

Recommended Resources

Bates, J. C., MacCarthy, F., & Warren-Riley, S. (2018). Emphasizing embodiment, intersectionality, and access: Social justice through technofeminism past, present, and future. *Computers and Composition Online.* http://cconlinejournal.org/techfem_si/03_Bates_Macarthy_Warren_Riley/

Chaterdon, K., & Silvester, K. (2015). Mindful writer/embodied writer: Universal design and multimodal argument. *Kairos, 20*(1). http://kairos.technorhetoric.net/20.1/index.html

Kalin, J., & Frith, J. (2016). Wearing the city: Memory p(a)laces, smartphones, and the rhetorical invention of embodied space. *Rhetoric Society Quarterly, 46*(3), 222–235. https://doi.org/10.1080/02773945.2016.1171692

Mendelson, M. (1998). The rhetoric of embodiment. *Rhetoric Society Quarterly, 28*(4), 29–50. https://doi.org/10.1080/02773949809391129

Shapiro, A. (2020). 'Embodiments of the invention': Patents and urban diagrammatics in the smart city. *Convergence: The International Journal of New Media Technologies, 26*(4), 751–774. https://doi.org/10.1177/135485652094180

See also: Materiality, Posthuman Practice, Accessibility, Bias Toward Action

16. Environments

Jacob Craig

D esign and making take place in a range of physical places and digital spaces, often working in concert with one another. On screens, in makerspaces, at kitchen tables, and in garages, people are engaging in acts of production, and the keyword "environments" calls attention to the whereness of making and design. Complicated acts of making and design require environments that can support "extremely complex, contingent" processes that sprawl across different digital and physical spaces where information can be moved from one space to the next, taking new shape and finding new life through each movement (Johnson-Eilola, 2005, p. 62).

Fig. 16.1. Design takes into account the influences of place and space of making. Photo by Brandon Lee on Unsplash. https://bit.ly/keyword-fig-16–1

In addition to supporting acts of making, physical and digital environments also shape the processes of making. Most prominently, perhaps, is that all making and design requires an infrastructure consisting not of just machines, tools, hardware, and software but also localized conventions of practice, standards, and organizational structures (DeVoss, Cushman, & Grabill, 2005). These more social and cultural concerns shape both the processes of and expectations about effective making and design. Further, depending on the limitations of a particular environment, processes of making and design may also include "hacking" or disrupting planned physical environments to make them conducive to particular tasks (Walls, Schopieray, & DeVoss, 2009).

Digital environments likewise shape acts of making and design because they are spaces designed for particular users and uses. As described by Anne Frances Wysocki and Julia I. Jasken (2004), "The design of software is thus also the design of users" (p. 35). If a user is imagined as being unimaginative or uninterested—"dull and uninventive"—by the designer of a digital environment, that user embodies that positionality to their designing and making in order to use the digital environment to complete their task (Wysocki & Jasken, 2004, p. 34). More specifically, the goals and values of the designers of digital environments profoundly shape what users *can do* and who users *can be* in particular environments. For instance, Bill Gates' vision of making technologies ubiquitous access points for information that pattern after existing technologies like televisions has led to a culture that readily adopts new technologies in every part of life, and those technologies are largely used to access and consume information (McCorkle, 2012, p. 182).

At some level, makers and designers do recognize that physical and digital environments shape their processes and products, and this recognition often occurs in the registers of their attention and persistence. Makers and designers make choices about what physical environment to inhabit and what digital environments to display based on how those environments benefit their processes (Pigg, 2014). Particular aspects of makers' and designers' environments (objects, devices, software, furniture, other people) come to hold particular importance for writers, leading to a ritualization or habituation of their processes within those environments (Rule, 2018). Over time, those writers continue to seek or create the beneficial aspects of past environments, and this influence of the past likewise continues to shape who makers and designers are and what they do (Craig, 2019).

References

Craig, J. (2019). Affective materialities: Places, technologies, and development of writing Processes. *Composition Forum, 41.* http://composition-forum.com/issue/41/affective-materialities.php

DeVoss, D. N., Cushman, E., & Grabill, J. (2005). Infrastructure and composing: The when of new-media writing. *College Composition and Communication, 57*(1), 14–44.

Johnson-Eilola, J. (2005). *Datacloud: Toward a new theory of online work*. Hampton Press.

Pigg, S. (2014). Emplacing mobile composing habits: A study of academic writing in networked social spaces. *College Composition and Communication, 66*(2), 250–275.

Rule, H. (2018). Writing's rooms. *College Composition and Communication, 69*(3), 402–432.

Walls, D. M., Schopieray, S., & DeVoss, D. N. (2009). Hacking spaces: Place as interface. *Computers and Composition, 26*(4), 269–287. doi:10.1016/j.compcom.2009.09.003

Wysocki, A. F., & Jasken, J. I. (2004). What should be an unforgettable face . . . *Computers and Composition, 21*(1), 29–48.

Recommended Resources

Alexis, C. (2016). The material culture of writing: Objects, habitats, and identities in practice. In S. Barnett & C. Boyle (Eds.), *Rhetoric through everyday things* (pp. 83–95). University of Alabama Press.

Brodkey, L. (1984). Modernism and the scenes(s) of writing. *College English, 49*(4), 396–418. https://doi.org/10.2307/377850

Casey, E. (1996). How to get from space to place in a fairly short stretch of time: phenomenological prolegomena. In S. Feld & K. Basso (Eds.), *Senses of Place* (pp. 13–52). School of American Research Press.

Mauk, J. (2003). Location, location, location: The "real" (e)states of being, writing, and thinking in composition. *College English, 65*(4), 368–388. https://doi.org/10.2307/3594240

Pigg, S. (2020). *Transient literacies in action: Composing with the mobile surround*. The WAC Clearinghouse/University Press of Colorado. https://wac.colostate.edu/books/writing/transient/

Rickert, T. (2013). *Ambient rhetoric: The attunements of rhetorical being*. University of Pittsburgh Press.

See also: Local, Materiality, Hacking, Cultural Intelligence, Community Engagement

17. Ethics

Steve Holmes

All designers and makers engage in making ethical choices even if they don't see their actions as particularly ethical. Makers choose sustainable materials and support local businesses or they source projects from large multinational chains with overseas factories. Designers take extra time to build apps with the need for accessibility in mind or they make an (all too often) expedient choice to limit their audience to able-bodied users. Let's do a mental inventory—have you taken the time to put your own spin on a recent design or making project instead of copying an existing design with no or minimal creative re-envisioning? Does your company use commercial design products or try to support open educational resource development?

While many think of ethics in terms of "right and wrong" conduct, ethics actually has to do with the broad types of reasoning that help us form a wide variety of ethical and unethical *values* that determine how we should act or think about a given situation or circumstance. Ethical matters also include questions of how to live a good life, be an empathetic friend, or enact ethical caregiving to children. One important reason to study ethics lies in being able to reflect more on the reasons behind why you believe and think about what you do as well as to start to realize some of the complexities or even unintended consequences behind certain ethical values you hold.

The Western philosophical traditions subdivides ethics into to three majors ethical paradigms:

- Deontology (rule-based; university)
- Consequentialism (ends justify the means)
- Virtue (character-based ethics) & Care (giving and receiving care)

It is critical to note that each framework offers different affordances and constraints. At times, you might benefit from a deontological approach, such as Immanuel Kant's "categorical imperative." Kant has argued that whatever we believe is an ethical action has to be ethical for any actor in any circumstance. Deontology is useful for theorizing copyright issues in design, for example, in order to protect makers' intellectual property in commercial projects. Of course, we cannot desire as a universal ethical action that we should be able to steal one another's property in all circumstances. Yet, other ethical dilemmas in copyright require virtue ethics, which is a flexible character-based form of prudential and pragmatic reasoning. Virtue ethics sees ethical action changing based on what the circumstances demand. When large companies like John Deere or Apple try to make their copyrighted software immune from tinkering to increase their revenue, a variety of Right to Repair movements have argued that it should be a case-by-case basis ethically good for makers to be able to interact with the software systems for devices that users own.

While Eurocentric attitudes have shaped how ethics was taken up in technical communication contexts, ethical frameworks have always been articulated across different cultures. In virtue ethics, Confucius offers alternative ways of theorizing individual honesty disclosure beyond Aristotle (Dragga, 1999; Wang & Gu, 2022). The Apache-Jarilla philosopher Viola Cordova (2004) similarly shifts Aristotle's focus on the autonomous individual (the "I") to an indigenous "We" to theorize virtue ethics from an indigenous perspective (Itchuaqiyaq, 2021). Similarly, the *Ubuntu/botho* tradition of moral reasoning from Sub-Saharan African moral argues that "A person is a person through other persons" (Metz & Gaie, 2010, p. 274). Different ethical frameworks give us different ethical values. For example, Kant's belief in essential human dignity supports patient-doctor privacy whereas *Ubuntu* tends to view any personal ethical decisions about care as a process that one's entire family or network of close associates should be intimately involved in.

Ethics is a particularly important area for design and making conversations. With the "critical" part of "critical making" conversations stemming from Matt Ratto's (2011) pioneering work, some makers refuse to see making or design as a set of value-neutral skills or techniques. They point to how certain forms of making have been valued (coding as white male dominated pursuits) and other forms of making or identities of makers (female, queer, non-white) have been devalued. In response, scholars who study critical making or social justice have been engaged in an ethical project called "critique." They rightly point to inequity in making and design practices (like "design justice") and argue for a reinclusion of excluded identities, cultural values or historically and multiply marginalized peoples.

Yet, ethical considerations for makers and designers must always include more than only the critique of inequality (Walwema et al., 2022). Corporate DEI (diversity, equity, and inclusivity) activists, for example, need to undertake self-care to cope with the stress of ongoing institutional pushback. Companies that sell sustainable products cannot be given a free pass for stiffing employees on parental leave for a newborn or failing to have a diverse upper-management team. Those who wish to remedy design inequality might find that they need to use limited forms of utilitarianism, for example, to identify solutions within a set of embodied, financial, material, and institutional constraints. Companies have sought to invest in DEI efforts recently because of utilitarian arguments as a case in point. Diverse workforces are more productive and help achieve the bottom line.

Yet, it is also true that many variations of utilitarianism can result in circumstances where a majority can feel ethically justified in oppressing a minority. For example, some companies may defend the cost-effectiveness of allocating most of their design budget for a new digital product to fulfill the needs of able-bodied users. Companies thereby might only allocate the bare minimum effort required by law to produce low-quality accessibility options because disabled users will not generate enough money to offset additional company resource allocation. Thus, designers and makers would need something like Kant's universal belief in human dignity and equality to steer some of utilitarianism's problems. The larger point is that ethical study will not give designers or makers a clear-cut list of approaches to enact in all moments of ethical decision making. Ethical study can help alert us to the wider range of consequences and possible alternatives to any given ethical value that we hold and seek to apply in our work.

References

Cordova, V. (2004). Ethics: The we and the I. In. A. Waters (Ed.), *American Indian thought: Philosophical essays* (pp. 173–181). Blackwell.

Dragga, S. (1999). Ethical intercultural technical communication: Looking through the lens of Confucian ethics. *Technical Communication Quarterly, 8*(4), 365–381.

Itchuaqiyaq, C. U. (2021). Iñupiat i itqusiat: An indigenist ethics approach for working with marginalized knowledges in technical Communication. In R. Walton & G. Agboka (Eds.), *Equipping technical communicators for social justice work: Theories, methodologies, and pedagogies* (pp. 33–38). University Press of Colorado.

Metz, T., & Gaie, J. B. (2010). The African ethic of *Ubuntu/Botho*: Implications for research on morality. *Journal of Moral Education, 39*(3), 273–290.

Ratto, M. (2011). Critical making: Conceptual and material studies in technology and social life. *The Information Society, 27*(4), 252–260.

Wang, X., & Gu, B. (2022). Ethical dimensions of app designs: A case study of photo- and video-editing apps. *Journal of Business and Technical Communication, 36*(3), 355–400.

Walwema, J., Colton, J. S., & Holmes, S. (2022). Introduction to special issue on 21st-century ethics in technical communication: Ethics and the social justice movement in technical and professional communication. *Journal of Business and Technical Communication, 36*(3), 257–269.

Recommended Resources

Colton, J. S., & Holmes, S. (2018). *Rhetoric, technology, and the virtues*. University Press of Colorado.

Colton, J. S., Holmes, S., & Walwema, J. (2017). From NoobGuides to #OpKKK: ethics of anonymous' tactical technical communication. *Technical Communication Quarterly, 26*(1), 59–75. https://doi.org/10.1080/10 572252.2016.1257743

Dragga, S. (2001). Ethics in technical communication [special issue]. *Technical Communication Quarterly, 10*(3), 245–249. https://doi.org/10.1207/ s15427625tcq1003_1

Katz, S. B. (1992). The ethic of expediency: Classical rhetoric, technology, and the holocaust. *College English, 54*(3), 255–275. https://www.jstor.org/ stable/378062

Markel, M. (2001). *Ethics in technical communication: A critique and synthesis*. Ablex Publishing.

Porter, J. E. (1993). Developing a postmodern ethics of rhetoric and composition. In T. Enos & S. C. Brown (Eds.), *Defining the new rhetorics* (pp. 207–226). Sage.

Ross, D. G. (2021). (Teaching) ethics and technical communication. In M. J. Klein (Ed.), *Effective teaching of technical communication* (pp. 67–87). WAC Clearinghouse.

See also: Advocacy, Feminist Making, Inclusive Design, Radical Imagination

18. Heuristics

Derek N. Mueller

Broadly construed, heuristics guide activity. Connotations of the term have been applied in design, rhetoric, psychology, education, and computation, in each disciplinary context bearing relationship to step-wise problem solving. Although heuristics may be explicit or implicit, functioning as a carefully documented set of instructions in the world (like, assembling a bicycle) or far more faintly as an observed and intuited imitation of another's step-wise approach (riding a bicycle), they function as rules that provide makers and designers with some degree of procedural regularity. Consider, for any of the following activities, the distinctive ways in which they are rule-governed, guided by instructions or steps that, when followed, elevate prospects for felicitous execution: parallel parking on a hill, weaving a pine needle basket, customizing a document's line spacing in a word processor, preparing guacamole without cilantro, and bathing a medium-sized dog. Unguided, such activities can go quite badly; loosely guided by heuristics, however, they chance *working*, over and over again.

Tracing the etymology of heuristics is insightful particularly for its associations with the Greek *heuriskein* ("heuristic," 2019), with relates to finding or discovering. For invention, heuristics also bears relations to the etymology of heresy and heretic, which stem from the Greek *hairetikós* ("herético," 2019), or choice. Both conceptual forbearers share a resemblance, as well, with "eureka!", an exclamation corresponding with breakthrough, discovery, and surprise. These and other valences of heuristics were especially influential in Gregory Ulmer's (1994) *Heuretics*, a theoretical monograph that elaborates an inventional method for writing with theory. Thus, the significance of heuristics is haloed by associations with invention, discovery, and change—a family of concepts as essential for rhetoric as they are for making and design thinking.

Wholly consistent with these etymological orbits involving finding, discovery, and choice, in *Invention in Rhetoric and Composition*, Janice Lauer (2004) points to debates in the mid-1980s invested in whether rhetorical invention was predominantly heuristic (genesis vis-à-vis making, doing, and experimentation) or hermeneutic (genesis vis-à-vis interpretation) (see Fig. 18.1).

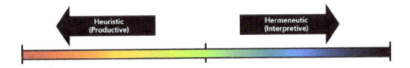

Fig. 18.1. A continuum indicating connection and distinction between heuristic and hermeneutic procedures. Source: Contributing author.

Lauer's treatment of heuristics in contexts of rhetorical invention set for it a disciplinary history while also positioning it sharply in contrastive relief, locating heuristics and hermeneutics along continua that distinguish them—even while linking them—in terms of production and interpretation, even if some contend to this day that hermeneutics (interpretation) are inventive. In addition to the tension suspended between heuristics (productive) and hermeneutics (interpretive), Lauer's extensive study of rhetorical invention draws upon Richard E. Young, Alton L. Becker, and Kenneth L. Pike's (1970) landmark volume, *Rhetoric: Discovery and Change*, which positions heuristic along a second continuum, this one shared by the terms aleatory (free-wheeling; open) and algorithm (strictly rule bound; closed) (see Fig. 18.2).

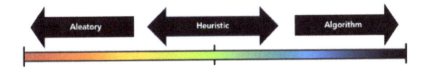

Fig. 18.2. A continuum indicating connection and distinction among aleatory, heuristic, and algorithmic procedures. Source: Contributing author.

Aleatory procedures are unbound by rules; algorithmic procedures, customarily associated with mechanical reproduction and computation, are heavily constrained, reproducible, and rule-bound. To position heuristics between these concepts acknowledges for it a generative capacity to guide making with greater or lesser strictures.

Making and design thinking offer abundant examples to plot along each continuum. Consider the example of greeting cards. On the first continuum, we can distinguish readily and also find connections between the steps involved in making a card (heuristic) and reading other cards for patterns or meaning (hermeneutic). Along the second continuum, imagine how differently the activities of making a greeting card would be guided by aleatory procedures and algorithmic procedures, between them variations that guide more or less loosely, granting choice while also stipulating gradually procedural conditions.

References

herético [Wiki]. (2019, June 4). Wiktionary. https://en.wiktionary.org/wiki/her%C3%A9tico

heuristic [Dictionary]. (2019, May 25). Oxford Dictionaries | English. https://en.oxforddictionaries.com/definition/heuristic

Lauer, J. M. (2004). *Invention in rhetoric and composition.* The WAC Clearinghouse.

Ulmer, G. L. (1994). *Heuretics: The logic of invention.* Johns Hopkins University Press.

Young, R. E., Becker, A. L., & Pike, K. L. (1970). *Rhetoric: Discovery and change.* Harcourt College.

Recommended Resources

Crowley, S. (1985). Invention in nineteenth-century rhetoric. *College Composition and Communication, 36*(1), 51–60. https://doi.org/10.2307/357606

Farrell, T. B. (1991). Practicing the arts of rhetoric: Tradition and invention. *Philosophy & Rhetoric, 24*(3), 183–212. https://www.jstor.org/stable/40237676

Lauer, J. M. (1967). *Invention in contemporary rhetoric: Heuristic procedures.* Doctoral dissertation. University of Michigan.

Muckelbauer, J. (2008). *The future of invention: Rhetoric, postmodernism, and the problem of change.* State University of New York Press.

Vitanza, V. J. (2000). From heuristic to aleatory procedures; or, toward "writing the accident." In M.D. Goggin (Ed.), *Inventing a discipline: Rhetoric scholarship in honor of Richard E. Young* (pp. 185–206). National Council of Teachers of English.

See also: Context & Contextual Design, User Requirements, Tools

19. Inter/Cross-Disciplinarity

Ann Shivers-McNair

Design, like any other profession, resides within specific disciplines. Disciplinarity organizes much of our work in academic and professional contexts. By defining an area of expertise, standards of practice, shared knowledge, and commonly held expectations, we can hold ourselves accountable, innovate, and work efficiently. But much of the work we do and the challenges we face involves multiple disciplines. Interdisciplinarity (or cross-disciplinarity) offers us a way to navigate work and challenges across disciplinary differences.

Frédéric Darbellay, Zoe Moody, and Todd Lubart (2017) emphasize that interdisciplinarity is not merely adding multiple disciplines together in ways that ultimately reinforce distinctions between disciplines. Such an approach might begin by assuming that one discipline has more expertise than another, or by comparing the differences among disciplines. Instead, interdisciplinarity "involves the collaboration and integration of specific disciplines in relation to a common object" (Darbellay, Moody, & Lubart 2017, p. xv). And, indeed, practices of making that are associated with the maker movement often embody this definition of interdisciplinarity by focusing on a problem or challenge instead of on a particular discipline. In this way, making can bring together an array of disciplines, from STEM (science, technology, engineering, and math) to arts and humanities, in service of solving problems.

As a learning experience in educational, professional, or community contexts, making can "provide multiple access points to learning and allow [people] to engage based on their interests first" (Regalla, 2016, p. 261). For example, a person attending a workshop at their local makerspace on programming and integrating LED lights into costume design might learn, draw upon, and integrate different disciplinary practices in the experience of making wearable technology for cosplay: from textiles, sewing, and color theory, to circuits, code, and soldering. Similarly, as a design strategy, an integrative, problem-focused approach to interdisciplinarity also connects with design thinking. Like practices of making associate with the maker movement, design thinking is "diffused through many disciplinary fields and practices," because the designer must aim "at the conception and solution of complex problems in an innovation perspective centered on human experiences which is not limited to disciplinary boundaries" (Darbellay, Moody, & Lubart 2017, p. xv). For example, an inventor working in a community makerspace may find a collaborative, interdisciplinary environment that is conducive to designing things that solve community problems or support community goals by focusing on human experiences rather than on a particular discipline or brand.

Such interdisciplinary practices offer the potential for what Irene Dölling and Sabine Hark (2000) describe as transdisciplinarity: the "continual reexamination of artificially drawn and contingent boundaries and that which they exclude" (p. 1197). In other words, by centering human/community experiences and challenges instead of disciplinary boundaries, practices of making and design can be both innovative and inclusive by working across differences. But inclusive interdisciplinarity (or transdisciplinarity) is not inherent in acts of making. Inclusive interdisciplinarity requires an intentional engagement across differences that is grounded in an ethical responsibility to the communities and environments for whom and in which we make.

References

Darbellay, F., Moody, Z., & Lubart, T. (2017). Introduction: Thinking creativity, design and interdisciplinarity in a changing world. In F. Darbellay, Z. Moody, and T. Lubart (Eds.), *Creativity, design thinking and interdisciplinarity* (pp. xi–xxii), Springer.

Dölling, I., & Hark, S. (2000). She who speaks shadow speaks truth: Transdisciplinarity in women's and gender studies. *Signs: Journal of Women in Culture and Society*, 25(4), 1195–1198.

Regalla, L. (2016). Developing a maker mindset. In K. Peppler, E. Halverson, & Y. B. Kafai, eds., *Makeology: Makerspaces as learning environments, Volume 1* (pp. 257–272). Routledge.

Recommended Resources

Blunden, A. (2010). *An interdisciplinary theory of action*. Brill.

Kleinberg, E. (2008). Interdisciplinary studies at a crossroads. *Liberal Education, 94*(1), 6–11. https://files.eric.ed.gov/fulltext/EJ790435.pdf

Moran, J. (2010). *Interdisciplinarity*. 2nd edition. Routledge.

Newell, W. H. (2001). A theory of interdisciplinary studies. *Issues in Integrative Studies, 19*(1), 1–25.

O'Rourke, M. O., & Crowley, S. J. (2013). Philosophical intervention and cross-disciplinary science: The story of the Toolbox Project. *Synthese, 190*, 1937–1954. https://doi.org/10.1007/s11229-012-0175-y

Pennington, D. D. (2008). Cross-disciplinary collaboration and learning. *Ecology and Society, 13*(2), 1–13. https://www.jstor.org/stable/26267958

See also: STEM/STEAM, Shared Leadership, Radical Collaboration

20. Invention

Erin Kathleen Bahl

How do we come up with something to make? Where do our ideas originate? How do we get started? Any designer who has ever stared at a blank page, empty screen, or blinking cursor knows the joys, frustrations, and uncertainties that accompany invention. "Invention" in making and design refers to processes of generating and developing ideas. Indeed, it may sometimes seem like a mysterious, almost magical process reserved for only a select few geniuses, or an elusive spark that can only be captured under just the right creative conditions. However, contemporary rhetorical approaches to invention emphasize the messy, material, and often collaborative nature of discovering new ideas.

Historically, invention is the classical rhetorical canon understood as "coming up with something to say," taking place at the beginning of the rhetorical act and initiating production (Aristotle, tr. 1984; Cicero, tr. 1984). This approach carried over to writing classrooms, in which invention exercises and heuristics were used to generate ideas that were then shaped into finalized texts (Lauer, 2004). However, recent invention scholarship informed by digital making emphasizes invention as an ongoing process that takes place throughout a developing artifact's lifecycle, in which ideas are continually generated and changing (Bahl, 2018; Buehl, 2016; Delagrange, 2009; Goggin, 2011). These approaches situate invention not just as a spark in the mind of the maker, writer, or designer, but rather as a process emerging from multiple material channels, collaborations, and experiences coming together in unique arrangements to create new meaning. This approach demystifies understandings of invention grounded in "the lonely writer in a garret," (Lunsford & Ede, 1983, p. 157) attending instead to the way ideas emerge out of particular, situated, often messy conditions, and continuously emerge and change throughout design processes.

When it comes to making and design, especially in digital spaces, scholars grounded in digital making suggest that invention involves a willingness to explore, get messy, play, fail, try again, and follow where the project leads, being open to new ideas as they emerge in the act of making (Andrews & Bentley, 2018; Garrett et al., 2012). Numerous factors may influence idea generation in making and design, including material conditions, social conditions, available media, collaborators, available publication infrastructures, influences and inspirations, aesthetic preferences and, of course, the maker/designer themselves, just to name a few.

One example of this approach to invention is exemplified in the *Inventio* section of *Kairos: A Journal of Rhetoric, Technology, and Pedagogy*, a peer-reviewed online scholarly journal that publishes webtexts, or scholarly works designed for an online environment. The *Inventio* section publishes webtexts that focus on the stories or processes behind how particular digital artifacts came to be, sharing their development practices from start to finish. These invention stories raise opportunities for critical questions on how an artifact might have turned out differently, as the designer discovered and invented its meaning along the way in negotiation with available material affordances and constraints. Stories such as these demystify artifacts (digital or otherwise) by making their development more transparent (Haas, 1996) and helping us understand a particular artifact's meaning in light of how it came to be.

References

Andrews, K., & Bentley, T. M. (2018). Multimodal composing, sketch-notes, and idea generation. *Kairos: A Journal of Rhetoric, Technology, and Pedagogy*, 22(2). http://kairos.technorhetoric.net/22.2/disputatio/andrews/index.html

Alexander, K. P., & Williams, D. M. (2015). DMAC after dark: Toward a theory of distributed invention. *Computers and Composition, 36*, 32–43.

Aristotle. (tr. 1984). *Rhetoric*, translated by W. Rhys Roberts. Random House.

Bahl, E. K. (2018). *Refracting webtexts: Invention and design in composing multimodal scholarship*. Doctoral dissertation. The Ohio State University.

Buehl, J. (2016). *Assembling arguments: Multimodal rhetoric and scientific discourse.* University of South Carolina Press.

Cicero, M. T. (tr. 1984). Rhetorical invention. *The orations of Marcus Tullius Cicero*, translated by C. D. Yonge (pp. 241–280). George Bell and Sons.

Delagrange, S. (2009). Wunderkammer, Cornell, and the visual canon of arrangement. *Kairos: A Journal of Rhetoric, Technology, and Pedagogy, 13*(2). http://kairos.technorhetoric.net/13.2/topoi/delagrange/

Garrett, B., Landrum-Geyer, D., & Palmeri, J. (2012). Re-inventing invention: A performance in three acts. In D. Journet, C. Ball, and R. Trauman (Eds.), *The new work of composing* (n.d.). Utah State University Press/Computers and Composition Digital Press. http://ccdigitalpress.org/nwc/chapters/garrett-et-al/

Goggin, M. D. (2011, 19 February). (Re)inventing inventio. Keynote Address for the 2011 Arizona State University Composition Conference. http://vimeo.com/20529630

Haas, C. (1996). *Writing technology: Studies on the materiality of literacy.* Lawrence Erlbaum Associates, Inc.

Lauer, J. (2004). *Invention in rhetoric and composition.* Parlor Press.

Lunsford, A., & Ede, L. (1983). Why write . . . together? *Rhetoric Review, 1*(2), 150–157.

Recommended Resources

Crowley, S. (1990). *The methodical memory: Invention in current-traditional rhetoric.* Southern Illinois University Press.

Gruwell, L. (2022). *Making matters: Craft, ethics, and new materialist rhetorics.* Utah State University Press.

Lauer, J. M. (2004). *Invention in rhetoric and composition.* Parlor Press.

LeFevre, K. B. (1987). *Invention as a social act.* Southern Illinois University Press.

Muckelbauer, J. (2008). *The future of invention: Rhetoric, postmodernism, and the problem of change.* State University of New York Press.

Ulmer, G. L. (1994). *Heuretics: The logic of invention.* Johns Hopkins University Press.

Warnick, B. (2000). Two systems of invention: The topics in the *Rhetoric* and *The New Rhetoric.* In A. G. Gross and A. E. Walzer (Eds.), *Rereading Aristotle's* Rhetoric (pp. 107–129). Southern Illinois University Press.

See also: Innovation, Bodystorming, Assemblage, Iterative Design

21. Learning Diversity

Michael Riendeau

As an educational exercise, design and its associated pedagogical activities need to recognize learning diversity. Since the 1960s, American schools and many others around the world have recognized a sizable subset of students as learning disabled. Under the Individuals with Disabilities Education Act of 2004 (IDEA), the most recent reauthorization of Pub. L. 94–142 (1975) and the federal law fundamentally responsible for the practices of our special education system, a specific learning disability is defined as:

> a disorder in one or more of the basic psychological processes involved in understanding or in using language, spoken or written, that may manifest itself in an imperfect ability to listen, think, speak, read, write, spell, or do mathematical calculations. This term includes such conditions as perceptual disabilities, brain injury, minimal brain dysfunction, dyslexia, and developmental aphasia. This term does not include children who have learning problems that are primarily the result of visual, hearing, or motor disabilities; mental retardation; or environmental, cultural or economic disadvantage (Individuals with Disabilities Education Act, 2004).

This understanding, often termed the medical model of learning disability, has prevailed in the field of special education since its inception. The need for a new, more equitable, and more practical way to understand the sometimes-unexpected failure of students in school is arguably one of the key issues of our time (Sleeter, 2010).

In the 1990s, critiques of the medical model of learning disabilities began to emerge in educational literature (see Skrtic, 1995). About twenty years ago, learning diversity was proposed as an alternative paradigm to the medical model (McDonald & Riendeau, 2002). Learning diversity is an educational approach that rejects the idea that learning disabilities are characteristic of individuals, understanding so-called "disabilities" as the intersection of rigid expectations and systems with the uniqueness of diverse learners.

Learning diversity (McDonald & Riendeau, 2003; Riendeau, 2015) is an approach to learning that privileges individualization, adhocracy, and emergence. In practice, learning diversity means beginning with and returning frequently to the aspirations, talents, and challenges of each learner and prioritizing those elements in the development of learning experiences. It means jettisoning predetermined curricula and methodologies and replacing them with a willingness to live in the less comfortable but more exciting space of the adjacent possible (Johnson, 2010).

Learning diversity complements makerspaces at schools and in communities. At the core of both is a desire for people to explore learning without being influenced by the pervasive bureaucracies on which education and industry are predicated. The maker movement, contends Dale Dougherty (2012), exists "because of people's need to engage passionately with objects in ways that make them more than just consumers" (p. 12). Learning diversity exists because there are educators who value the organic intellectual and social development of students as they creatively and intuitively explore the humanities unhindered by rigid standards. Both learning diversity and the maker movement create spaces for students and faculty who value solving problems together. The maker movement enthusiastically values doing-it-with-others. A learning diversity approach in the makerspace (and in all learning spaces) emphasizes adhocracy (Kim, 2019). Whereas the unintended consequences of organizational bureaucracies in schools transform teaching and learning from collaborative endeavors based on authentic human relationships to factory-model assembly lines where standard practices are imposed upon us and our students, adhocracy brings those collaborative endeavors back into the teaching and learning relationship.

Creating in a makerspace that is organized around a learning diversity approach also means privileging emergence, or being available to act on individual makers' needs in the moment. These needs often range from the mundane acts of providing makers instructions using alternative language or examples to the more creative acts of offering makers opportunities to create their own invention and problem-solving strategies to complete a task or project. Emergence also means inviting makers into the space with the understanding that they bring with them prior learning experiences, abilities, apprehensions, and creative ideas that inform their innovations. Individually or in a group, makers explore the path from idea to final product through product design, rhetorical strategies, production design, and manufacturing processes, and we know that along this path makers may enter into the adjacent possible, or what could be. In a makerspace that privileges emergence, we accept and expect that the path our makers travel is influenced by what Thomas Skrtic (1995) calls the "inevitability of human diversity" (p. 248), or the fact that each of us approaches thinking, writing, moving, and making differently. Perhaps above all, openly privileging emergence in the makerspace provides our makers the confidence they need in their ability to move forward with innovative ideas.

A learning diversity approach is distinguished from Universal Design for Learning (UDL) in its broader consideration of content, goals, and "standards" in addition to methodologies and modalities for learning. While UDL can accommodate a more thoroughgoing consideration of expected outcomes for individual students, it focuses primarily on multiple means of accessing a predetermined curriculum. Learning diversity theory requires an iterative teleological process for each student.

References

Dougherty, D. (2012). The maker movement. *Innovations, 7*(3), 11–14.

Johnson, S. (2010). *Where good ideas come from: The natural history of innovation.* Riverhead Books.

Individuals with Disabilities Education Act. (2004). *Public Law,* 108–446. https://sites.ed.gov/idea/statute-chapter-33.

Kim, M. (2019). Teaching and learning in the adhocratic school. Eagle Hill School Blog. https://www.eaglehill.school/blog-detail?pk=1013946&fromId=240404

McDonald, P., & Riendeau, M. (2003). A Copernican revolution in learning. *Independent School,* Winter edition.

Riendeau, M. (2015). *How schools created learning disabilities and what they can do about it.* Hardwick, MA: Eagle Hill School. https://bbk12e1-cdn.my-schoolcdn.com/ftpimages/190/misc/misc_161508.pdf

Skrtic, T. M. (1995). *Disability and democracy: Reconstructing (special) education for postmodernity.* Teachers College.

Sleeter, C. (2010). Why is there learning disabilities? A critical analysis of the birth of the field in its social context. *Disability Studies Quarterly, 30*(2), 210–237.

Recommended Resources

Capp, M. J. (2017). The effectiveness of universal design for learning: A meta-analysis of literature between 2013 and 2016. *International Journal of Inclusive Education, 21*(8), 791–807. https://doi.org/10.1080/13603116.2017.1325074

Center for Applied Special Technology. (2018). University design for learning guidelines (version 2.2). https://udlguidelines.cast.org/

Kumi-Yeboah, A., Yanghyun, K., Sallar, A. M., & Kiramba, L. K. (2020). Exploring the use of digital technologies from the perspective of diverse learners in online learning environments. *Online Learning, 24*(4), 42–63. https://doi.org/10.24059/olj.v24i4.2323

Parrish, N. (2019). Ensuring that instruction is inclusive for diverse learners. Edutopia. https://www.edutopia.org/article/ensuring-instruction-inclusive-diverse-learners

Sampson, R. J. (2020). The feeling classroom: Diversity of feelings in instructed L2 learning. *Innovation in Language Learning and Teaching, 14*(3), 203–217. https://doi.org/10.1080/17501229.2018.1553178

See also: Accessibility, Inclusion, Inclusive Design

22. Local

Madison Jones

Attention to location is an increasingly important consideration for design thinking in an emerging age of digital media. While global networks are often imagined through the ethereal metaphors of "the cloud" (Hu, 2016), digital communication takes place across virtual and actual networks (Reid, 2007) that rely on site-specific cultural and historical conditions (Starosielski, 2015) as well as material infrastructures, resources, and environments (Schivener & Edwards, 2020). In an era of global communication, it is important then, to recognize place as a central aspect of information design. While global approaches and large-scale issues often frame design methodologies, engaging location is key to successful communication. From locative technologies like augmented reality (or AR) and digital maps to the practice of localization in user-centered design, making "takes place" on location (Dobrin, 2001).

Fig. 22.1. Language diversity in the global marketplace. Photo by Nic Low on Unsplash. https://bit.ly/keyword-fig-22–1

Communication always happens somewhere, and we often take for granted how important *kairos* (a rhetorical term referring to the opportune time and place) is for successful information exchange (Miller, 1994; Rickert, 2007). The connections between place, invention, and memory are also deeply entwined in Aristotle's concept of *topos*, bringing together topography with topic when we "take a stance" or position an argument (Miller, 2000; Muckelbauer, 2008) through topologies (Walsh & Boyle, 2017). Making and invention occurs across digital and material interfaces, and good design practices require attentiveness to the contingent needs and values of specific audiences, situations, and locations (Shivers-McNair & San Diego, 2017). Over the past few decades, writing studies scholars have begun to address the important role that location plays in the process of invention and circulation (Reynolds, 2007). Place-based inventive practices focus on the recursive relations between digital information and place (Rice, 2012), treating these emerging effects as ecological (Edbauer, 2005; Jones 2021) and chaotic rather than static (Dobrin, 2007). Places shape, and are shaped by, design.

In user-centered design, places present discrete obstacles and affordances for cross-cultural communication (Gonzales & Zantger, 2015). User-localization, the practice of translating information for specific languages and cultures, offers important opportunities for design thinking to engage with the networked conditions of specific cultural practices (Jiménez-Crespo, 2010). From remix to digital activism, online writing environments likewise present complex and contested territories for design and circulation (Medina & Pimentel, 2018). While the availability of internet access often exacerbates inequity in marginalized populations, examples like the Equitable Internet Initiative in Detroit, MI demonstrate how local communities can counteract the structural inequalities presented by global telecom's control over network access (Greene & Jones, 2019). It is important, therefore, that designers engage communities in reciprocal dialog as they design (Powell and Takayoshi, 2003).

For mobile media like augmented reality (AR), location functions as more than a mere site upon which to inscribe. Scholars interested in the locative affordances of AR see the capabilities of emerging technologies for place-based thinking and public activism (Tinnell, 2017). For instance, Jacob Greene (2017) draws upon articulation theory to describe the relationships between the material environment and digital content, which he argues are not simply a static layering, but an emergent process of co-production. In another example, Roberta Chevrette (2016) discusses how "holographic rhetoric" enables a sense of place that brings "attention to contested memories as they unfold within the imaginative and material geographies of

US settler colonialism" (p. 150). These studies demonstrate the rich complexity that place often presents for makers.

When designing media for specific users (such as when producing mobile smartphone content), understanding *where* the user will access the information provides a host of considerations, from the limitations of the material environment (such as a bright, hot, or noisy street corner) to the kinds of activities users may likely be engaged in (such as museum visitors scanning a QR code for information about an exhibit or commuters listening to a podcast). Moreover, attention to the locations of writing means engaging the complex ecologies of place, the circulation of virtual and material networks, the histories and cultures of audiences and users, and the limitations and affordances of the material environment; these name but a few location-based considerations for makers and designers.

References

Dobrin, S. (2007). *Postcomposition*. Southern Illinois University Press.

Dobrin, S. (2001). Writing takes place. In C. Weisser and S. Dobrin (Eds.), *Ecocomposition: Theoretical and pedagogical approaches* (pp. 11–25). State University of New York Press.

Edbauer, J. (2005). Unframing models of public distribution: From rhetorical situation to rhetorical ecologies. *Rhetoric Society Quarterly*, *35*(4), 5–24.

Gonzales, L., & Zantger, R. (2015). Translation as a user-localization practice. *Technical Communication*, *62*(4), 271–284.

Greene, J. (2017). From Augmentation to articulation: (Hyper)linking the locations of public writing. *Enculturation*, *24*. http://enculturation.net/from_augmentation_to_articulation

Greene, J., & Jones, M. (2019). Articulate Detroit: Visualizing environments with augmented reality. *Computers and Composition Online*. http://cconlinejournal.org/articulatedetroit/

Jiménez-Crespo, M. (2010). The intersection of localization and translation. *Translation and Interpreting Studies*, *5*(2) 186–207.

Jones, M. (2021). A counterhistory of rhetorical ecologies. *Rhetoric Society Quarterly*, *51*(4), 336–352.

Hu, T. (2016). *A Prehistory of the cloud*. The MIT Press.

Medina, C., & Pimentel, O. (2018). *Racial shorthand: Coded discrimination contested in social media*. Logan, UT: Utah State University Press. https://ccdigitalpress.org/book/shorthand/

Miller, C. (1994). Opportunity, opportunism, and progress: *Kairos* in the rhetoric of technology. *Argumentation*, *8*(1), 81–96.

Miller, C. (2000). The Aristotelian topos: Hunting for novelty. In A. G. Gross and A. E. Walzer (Eds.), *Rereading Aristotle's* Rhetoric (pp. 130–146). Southern Illinois University Press.

Muckelbauer, J. (2008). *The future of invention: Rhetoric, postmodernism, and the problem of change*. New York, NY: State University of New York Press.

Powell, K. and Takayoshi, P. (2003). Accepting roles created for us: The ethics of reciprocity. *CCC*, 54(3), 394–422.

Reid, A. (2007). *The two virtuals: New media and composition*. Parlor Press.

Reynolds, N. (2007). *Geographies of writing: Inhabiting places and encountering difference*. Southern Illinois University Press.

Rice, J. (2012). *Digital Detroit: Rhetoric and space in the age of the network*. Southern Illinois University Press.

Rickert, T. (2007). Invention in the wild: On locating *kairos* in space-time. In. C. Keller and C. Weisser (Eds.), *The locations of composition* (pp. 71–89). State University of New York Press.

Shivener, R. & Edwards, D. (2020). The environmental unconscious of digital composing: Mapping climate change rhetorics in data center ecologies. *Enculturation, 32*. https://www.enculturation.net/environmental_unconscious

Shivers-McNair, A., & San Diego, C. (2017). Localizing communities, goals, communication and inclusion: A collaborative approach. *Technical Communication, 64*(2), 97–112.

Starosielski, N. (2015). *The undersea network*. Durham, NC: Duke University Press.

Tinnell, J. (2017). *Actionable media: Digital communication beyond the desktop*. Oxford University Press.

Walsh, L. & Boyle, C. (Eds.). (2017). *Topologies as techniques for a post-critical rhetoric*. Palgrave.

Recommended Resources

Chaput, C., & Hanan, J. S. (2019). Rhetorical hegemony: Transactional ontologies and the reinvention of material infrastructures. *Philosophy & Rhetoric, 52*(4), 339–366.

Dobrin, S. (2012). *Ecology, writing theory, and new media: Writing ecology*. Routledge.

Jensen, T. (2019). *Ecologies of guilt in environmental rhetorics*. Springer.

McKinnon, S., Asen, R., Chávez, K., & Howard, R. (2016). *Text + field: Innovations in rhetorical method*. Penn State University Press.

Rai, C., & Druschke, C. (2018). *Field rhetoric: Ethnography, ecology, and engagement in the places of persuasion*. University of Alabama Press.

Taffel, S. (2019). *Digital media ecologies: Entanglements of content, code and hardware.* Bloomsbury Academic.

Zakrzewski, P. (2022). *Designing XR: A rhetorical design perspective for the ecology of human+computer systems.* Emerald Publishing Limited.

See also: Environments, User Requirements, Context & Contextual Inquiry, Fieldwork

23. Maker Culture

Johansen Quijano

Although DIY (do-it-yourself) practices and DIY ethic in America can be traced back to the early 1900s with the launch of *Popular Mechanics* (Seelhorst, 1993), its most recent iteration—maker culture—finds its origins in *Make:* magazine, a bimonthly magazine launched in January 2005 that shares and discusses DIY projects on robotics, wood and metal working, and electronics. At its core, maker culture is a series of social norms, ideals, behaviors, and practices that embody an ethos of creativity, hacking, tinkering, and the re-envisioning and repurposing of already-existing artifacts. It exists at the intersection of technology-based DIY culture, hacker culture, and creative culture. It's a culture based on the principles that Mark Hatch discusses in *The Maker Movement Manifesto*: that because "making is fundamental to what it means to be human" one must "learn how to make" while being playful with what is being made (2014, p. 2).

Because its philosophy revolves around learning by doing in social environments and sharing what is made, maker culture encourages social active learning, informal learning, peer-led apprenticeships, and out-of-the-box thinking. This approach to collaborative DIY making allows maker culture to be "a bridge to pull communities back into sharing and face-to-face interaction as people help one another" (Inventionland, 2018). This kind of maker-based thinking gave rise to makerspaces, fablabs, techshops, and hackerspaces (Cavalcanti, 2013). Maker culture as a way of thinking and learning has expanded beyond its initial audience, and can now be seen at play in any of the many makerspaces in the US and all across the world (Lou & Peek, 2016) and in events like Nashville's Modern Maker Expo and Fort Worth's FabNOW Maker Conference.

A recent example of a maker culture mindset in practice can be seen in the story of Logan Moore (Sparks, 2019). Logan Moore, a two-year-old from a small town in Georgia, suffers from hypotonia, a condition that weakens a person's muscles and prevents them from being able to walk on their own. In many children's cases, this is treated with a special walker; however, Logan's parents were unsure as to whether the insurance would pay for his walker. According to reports, Logan's parents went to the local hardware store to buy some PVC pipes and make a DIY child-sized walker (Pelletiere, 2019). After helping Logan's parents to find the required parts, the store staff built Logan a PVC walker using tape, a few power tools, a bit of ingenuity, and the maker culture ethos that prompts people into action.

References

Cavalcanti, G. (2013, May 22). Is it a hackerspace, makerspace, techshop, or fablab? *Make:* https://makezine.com/2013/05/22/the-difference-between-hackerspaces-makerspaces-techshops-and-fablabs/

Hatch, M. (2014). *The maker movement manifesto: Rules for innovation in the new world of crafters, hackers, and tinkerers.* McGraw-Hill Education.

Inventionland. (2018, September 21). Makers & maker movements: What is maker culture? https://inventionland.com/blog/makers-maker-movements-what-is-maker-culture/

Lou, N., & Peek, K. (2016, February 23). By the numbers: The rise of the makerspace. https://www.popsci.com/rise-makerspace-by-numbers

Pelletiere, N. (2019, May 29). 2-year-old has sweetest reaction to new walker built from PVC piping. *Good Morning America.* https://www.goodmorningamerica.com/living/story/year-sweetest-reaction-walker-built-home-depot-employees-63319367

Seelhorst, M. (1993, March). Ninety years of popular mechanics. Possible dreams: Enthusiasm for technology in America. *Popular Mechanics, 170*(3), 46–83.

Sparks, H. (2019, May 30). Home Depot heroes build walker for toddler with muscle condition. *New York Post.* https://nypost.com/2019/05/30/home-depot-heroes-build-walker-for-toddler-with-muscle-condition/

Recommended Resources

Craig, W. (2018, July 06). What is "maker culture," and how can you put it to work? *Forbes.* https://www.forbes.com/sites/williamcraig/2015/02/27/what-is-maker-culture-and-how-can-you-put-it-to-work/#5cf06932540b

Footer, G., & Verhoeven, E. (2019). Tactics for a more-than-human maker culture. In L. Bogers and L. Chiappini (Eds.), *The critical makers reader: (Un)learning technology* (pp. 72–85). Institute of New Cultures, Amsterdam University of Applied Sciences.

Kohtala, C., Boeva, Y., & Troxler, P. (2020). Alternative histories in DIY cultures and maker utopias. *Digital Culture & Society, 6*(1), 5–34. https://doi.org/10.14361/dcs-2020-0102

Muir, A. (2019, January). Modern maker expo. https://www.facebook.com/modernmakerexpos/

Toups, D. (2019, April 11). Fab Now Maker Conference. https://www.tccd.
   edu/community/conferences-and-seminars/fab-now/

See also: Maker Movement, Makerspace, Feminist Making, Fablab

24. Maker Movement

Johansen Quijano

B uilding on the principles of a budding maker culture, the maker movement finds its roots in the maker faire, an event pioneered by Dale Dougherty in 2006 for makers, tinkers, crafters, and educators to showcase what they have made throughout the year and share what they have learned with other makers. This inclusive movement proposes that "we are all makers: as cooks preparing food for our families, as gardeners, and as knitters" (Dougherty, 2012, p. 11) The movement is about reclaiming what Dougherty calls a "tinkering spirit" whereby one could fix their own car, improve their own home, or make their own clothes; a spirit that, Dougherty suggests, has been lost over the last few decades.

Spurred by a community of hobbyists and enthusiasts, the maker movement expanded from the 2006 Bay Area Maker Faire. In 2007 two flagship maker faires were held, one in San Mateo, California, and the other in Austin, Texas. During the next few years, maker faires of all sizes were held throughout the world, with the official maker faire catalog listing over one thousand successful events. Furthermore, other maker events that are not necessarily affiliated with the maker faire take place year-round all over the world.

The maker movement has become so widespread that it has transcended its original audience of tinkerers and makers and can now be seen revitalizing business and education practices. In business, the maker movement has helped "shorten the time and cost from idea to prototype, provide networking with strategic partners, meet new consumer demand for customization, and breed success through people diversity" (Zwilling, 2017, n.p.). Meanwhile, in classrooms teachers are adopting practices that stem from the maker movement to get students interested in design thinking and STEM fields, while administrators are looking at how they can adapt maker spaces into learning commons spaces (Davis, 2014).

An example of the maker movement in action can be seen in the yearly FabNOW and MakerCon events in Fort Worth, Texas (Toups, 2019). In this event, makers from the community get together in a weekend-long expo where they showcase what they have made and share what they have learned. Workshops and discussions on 3D printing, sewing, and other maker practices are offered at different points during the event, and members of the local maker community teach lessons on making with various tools. The event also serves as a hub for businesses and schools to learn more about the maker movement and to get into the spirit of making.

References

Davis, V. (2014, July 18). How the maker movement is moving into classrooms. https://www.edutopia.org/blog/maker-movement-moving-into-classrooms-vicki-davis

Dougherty, D. (2012). The maker movement. *Innovations: Technology, Governance, Globalization, 7*(3), 11–14. doi:10.1162/inov_a_00135

Toups, D. (2019, April 11). Fab Now Maker Conference. https://www.tccd.edu/community/conferences-and-seminars/fab-now/

Zwilling, M. (2019, April 17). 8 ways the maker movement turns ideas into businesses. https://smallbizclub.com/startup/making-your-business-official/8-ways-maker-movement-turns-ideas-businesses/

Recommended Resources

Bajarin, T. (2014, May 19). Maker faire: Why the maker movement is important to America's future. http://time.com/104210/maker-faire-maker-movement/

Craig, W. (2015, February 27). What is "maker culture," and how can you put it to work? *Forbes.* https://www.forbes.com/sites/williamcraig/2015/02/27/what-is-maker-culture-and-how-can-you-put-it-to-work/

Fernandez, C. (2015, May/June). The origins of the maker movement. https://www.bbvaopenmind.com/en/technology/innovation/the-origins-of-the-maker-movement/

Grace-Flood, L. (2017, October 31). Open world: 20 databases to help you find local maker resources. *Make:.* https://makezine.com/2017/11/01/open-world-20-databases-help-find-local-maker-resources/

Hatch, M. (2014). *The maker movement manifesto: Rules for innovation in the new world of crafters, hackers, and tinkerers.* McGraw-Hill Education.

Inventionland. (2018, September 21). Makers & maker movements: What is maker culture? https://inventionland.com/blog/makers-maker-movements-what-is-maker-culture/

Make: Community. (n.d.). About maker faire. https://makerfaire.com/makerfairehistory/

Maker faire catalog: Find a faire near you. (2019, May 31). https://maker-faire.com/globalmap/

Morin, B. (2013, July 02). What is the maker movement and why should you care? *Huffington Post.* https://www.huffpost.com/entry/what-is-the-maker-movemen_b_3201977

See also: Maker Culture, Makerspace, Feminist Making, Physical Computing

25. Materiality

Jason Tham

Thanks to advances in human-computer interaction (HCI) technologies, greater attention has been given to multisensory (with emphasis on haptic and kinesthetic) experiences in composing—leading to new scholarship on embodiment and materiality in multimodal composition (Haas & Witte, 2001; Arola & Wysocki, 2012; Rifenburg, 2014; Rhodes & Alexander, 2014). In multimodality and design theories, materiality tends to refer to the observation of modes being taken to be the product of a maker shaping physical materials into meaningful artifacts. Matthew Davis and Kathleen Yancey (2014), in their discussing the role of materiality on assessment of multimodal texts, cite Lester Faigley (1999) for his argument about modality and materiality in multimodality:

> Images and words have long coexisted on the printed page and in manuscripts, but relatively few people possessed the resources to exploit the rhetorical potential of images combined with words. My argument is that literacy has **always** been a material, multimedia construct but we only now are becoming aware of this multidimensionality and materiality because computer technologies have made it possible for many people to produce and publish multimedia presentations. (p. 175, emphasis original)

Christina Haas (1996) has particularly pointed out the material dimensions of literacy and writing, with the term "material" referring to anything that possesses mass or matter, and which uses physical space. For multimodality, this includes any tools or resources that cross between the composer and his or her artifacts. In this sense, the material elements of the composing space—the pencils, desks, chairs, screens, keyboards, and other literacy materials—function as heuristics for learning. The connections between materials, users/composers, and the literacy knowledge in the composing environment are often mapped onto the socio-material conditions of learning as a way of problematizing their relations to the wider societal issues. In rhetoric and writing studies, scholars have been increasingly relying on activity and circulation theories to study the mediating power of tools as tied to knowledge making and dissemination (refer to Prior & Shipka, 2003; Trimbur, 2000). These socio-cultural and historical approaches to composing and multimodality emphasize the active and dynamic role of tangible materials, and the vitality of their interplay with learning and making.

In "Polymorphous Perversity in Texts," Johndan Johnson-Eilola (2012) makes an argument that multimodal theories can be expanded by seeing multimodal texts as multidimensional texts—beyond just signs and symbols. Johnson-Eilola challenges us to think about how we take pleasure in texts by interacting with them through fragmentation, unmaking, and remaking:

> I want to ask what happens when we begin to take less-authorized, polymorphously perverse pleasure in our texts, when we begin to treat texts less as objects out there and more as objects that we—literally—transgress the boundaries of, fragment, unmake, and remake. (Johnson-Eilola, 2012, n.p.)

Fig. 25.1. A composition of things. Photo by Luca Laurence on Unsplash. https://bit.ly/keyword-fig-25–1

Alluding to maker culture, Johnson-Eilola highlights the importance of remix/remake in text ownership: "If you can re-make an object, you don't really own it" (2012, n.p.). Jim Brown and Nathaniel Rivers (2013) also envision an object-oriented future for writing studies, one where students compose objects like puzzles and glass sculptures with ads and packaging for their objects. This future that Brown and Rivers imagine is partially an extension of Shipka's (2011, 2013) multimodal composition theory and partially an enactment of Ian Bogost's (2012) call to include all matter and not just "written matter" in humanities scholarship. As David Sheridan (2010) writes, "three-dimensional objects do indeed function rhetorically and may even possess their own distinctive rhetorical power. In fact, three-dimensional objects appear to play a unique role in fashioning culture itself" (p. 255).

Evidently, making as a practice challenges us to reconsider the viability of invention and delivery in an age of rapid innovation. For the purpose of advancing technical communication, we can study how prototyping changes the way we traditionally think of creation and final products. Failures and incompletions are common occurrences in the makerspaces; how do they help us rethink creativity? How might that affect our practices in the workplace and technical communicators' collaboration with designers and developers? These are the questions we should continue to explore in our teaching and research.

References

Arola, K., & Wysocki, A. F. (2012). *Composing(media) = composing(embodiment)*. Utah State University Press.

Bogost, I. (2012). *Alien phenomenology, or, what it's like to be a thing*. University of Minnesota Press.

Brown, J., & Rivers, N. (2013). Composing the carpenter's workshop. *O-Zone: A Journal of Object-Oriented Studies, 1*(1), 27–36.

Davis, M., & Yancey, K. (2014). Notes toward the role of materiality in composing, reviewing, and assessing multimodal texts. *Computers and Composition, 31,* 13–28.

Faigley, L. (1999). Material literacy and visual design. In J. Selzer, and S. Crowley (Eds.), *Rhetorical bodies: Toward a material rhetoric* (pp. 171–201). University of Wisconsin Press.

Haas, C. (1996). *Writing technology: Studies on the materiality of literacy*. Lawrence Erlbaum Associates.

Haas, C., & Witte, S. (2001). Writing as embodied practice: The case of engineering standards. *Journal of Business and Technical Communication, 15,* 413–457.

Johnson-Eilola, J. (2012). Polymorphous perversity in texts. *Kairos, 16*(3). http://kairos.technorhetoric.net/16.3/topoi/johnson-eilola/index.html

Prior, P., & Shipka, J. (2003). Chronotopic laminations: Tracing the contours of literate activity. In C. Bazerman and D. Russell (Eds.),

Writing selves, writing societies (180–238), Fort Collins, CO : The WAC Clearinghouse. http://wac.colostate.edu/books/selves_societies/prior/

Rhodes, J., & Alexander, J. (2014). *On multimodality: New media in composition studies*. National Council of Teachers of English.

Rifenburg, M. (2014). Writing as embodied, college football plays as embodied: Extracurricular multimodal composing. *Composition Forum*, 29. http://compositionforum.com/issue/29/writing-as-embodied.php

Sheridan, D. (2010). Fabricating consent: Three-dimensional objects as rhetorical compositions. *Computers and Composition, 27*(4), 249–265.

Shipka, J. (2011). *Toward a composition made whole*. University of Pittsburgh Press.

Shipka, J. (2013). Including, but not limited to, the digital: Composing multimodal texts. In T. Bowen and C. Whithaus (Eds.), *Multimodal literacies and emerging genres* (pp. 73–89). University of Pittsburgh Press.

Trimbur, J. (2000). Composition and the circulation of writing. *College Composition and Communication, 52*(2), 188–219.

Recommended Resources

Barnett, S., & Boyle, C. (Eds.). (2016). *Rhetoric through everyday things*. University of Alabama Press.

Coole, D., & Frost, S. (Eds.). (2010). *New materialisms: Ontology, agency, and politics*. Duke University Press.

Ricket, T. (2013). *Ambient rhetoric: The attunements of rhetorical being*. University of Pittsburgh Press.

Piña, M. (2020). Who cares if Johnny writes with a pencil? Or, a hauntological historiography of materiality in composition-rhetoric. *Rhetoric Review, 39*(2), 188–201. https://doi.org/10.1080/07350198.2020.1727079

Scollon, R., & Scollon, S. (2003). *Discourses in place: Language in the material world*. Routledge.

See also: Embodiment, Multigenre, Feminist Making, Physical Computing

26. Memory

Joe Cirio

Memory describes our representations of the past. Design thinking relies upon memory—and how we perceive and invoke the past—as a point of departure in the making and creation of text. Design concepts like *assemblage* and *remix* require a creator's knowledge of the meaning—or memory—audiences attach to antecedent texts to create something new from those texts (e.g., McElroy, 2015). *Intertextuality* describes how all texts are inhabited by and have a relationship to the texts that came before (Porter, 1986). Likewise, *genre* describes a texts' relationship to recurring rhetorical situations (Devitt, 1993)—identifying a genre gives meaning to a situation because a genre invokes previous rhetorical situations where that genre was used. Knowing the past, invoking the past, and representing the past are processes of memory. Indeed, the substance of our knowledge is grounded in memory: we cannot read and understand a text without memories attached to its parts, and we cannot create new texts without memory of what texts can invoke for us.

Conceptually, memory has been conceived as a place we've come from or a space we construct and inhabit. bell hooks (1989), reflecting on the experience of writing, explains that one must "incorporate . . . a sense of place, of not just who I am in the present but where I am coming from, the multiple voices within me . . . when I say then that these words emerge from suffering, I refer to that personal struggle to name the location from which I came to voice" (p. 16). hooks understood memory as a place we come from which defines where we are headed. hooks' conception of memory is reminiscent of theorists like Cicero and Quintillian who, likewise, thought of memory in terms of a space we can construct and visualize. Namely, we build "memory palaces" or visual spaces in the mind's eye and fill its corridors with images, objects, and people you'd need to recall later. In such configurations, memory was understood primarily as an interior process of individuals to visualize what came before (see Block, 2007). However, such representations of the past are not limited to mental images—rather, memory is also mediated by objects we place in the literal spaces we inhabit.

The ways in which we externalize memory are familiar for many of us: writing in a diary, making a scrapbook or photo album, scratching your initials in wet cement each describe a kind of memory work. In these instances, these material objects represent and distill moments in the past, and thus, they mediate memory. Like any mediated activity, memory is rhetorical and prompts questions about what kind of memory we construct, how, for whom, and for what purpose. Addressing these questions has been particularly relevant in the wake of recent tensions over the placement and reverence of confederate memorials in public spaces. Questions of making memory, and what gets to count in making our collective memory, continue to be a fraught and high stakes area.

But such questions can also be raised in a workspace. Aside from monuments, memory is also augmented through tools, technologies, objects, and materials that extend the capabilities of human memory. For instance, I think about the writing environments I made and inhabited while writing my dissertation: sitting at my desk in my apartment, I relied on my laptop computer's memory storage to access previous drafts and PDFs of articles I've collected. I often had books or printed articles in arm's reach, within which, over the years, I've inserted tabs and underlined key passages for future use. When writing in my shared office space with other graduate students, I often engaged in conversation about other research and ideas I hadn't considered. Such workspaces function as—what Johndon Johnson-Eilola (2005) calls—dataclouds: a "shifting and only slightly contingently structured information space" that's structured around tools and technologies that each serve a function in the activities of composition (p. 4). In the datacloud, I rely on the system of tools, objects, materials, and people that occupy my workspace in order to access information that my natural human memory cannot. Thus, as Paul Prior et al. (2007) and Derek Van Ittersum (2009) have observed, memory operates as distributed cognition where writers and makers access and mobilize the information from externalized memory systems, whether from online search engines, computer file storage, print documents with margin notes, or one's own brain. But like a memorial in public space, the objects and technologies in a datacloud that augment and enhance human memory can also define the bandwidth of what's possible by defining how that memory is accessed and thus, selecting what can be remembered.

References

Block, D. (2007). *Aristotle on memory and recollection: Text, translation, interpretation, and reception in western scholasticism.* Brill.

hooks, b. (1989). Choosing the margins as space of radical openness. *Frameworks, 36,* 15–23.

Devitt, A. (1993). Generalizing about genre: A new conception of an old concept. *College Composition and Communication, 44(4),* 573–586.

Johnson-Eilola, J. (2005). *Datacloud: Toward a new theory of online work*. Hampton Press.

McElroy, S. (2015). Assemblage by design: The postcards of Curt Teich and Company. *Computers and Composition, 37*, 147–165.

Porter, J. (1986). Intertextuality and the discourse community. *Rhetoric Review, 5(1)*. 34–47.

Prior, P., Solberg, J., Berry, P., Bellwoar, H., Chewning, B., Lunsford, K. J., Rohan, L., Roozen, K., Sheridan-Rabideau, M. P., Shipka, J., Van Ittersum, D., & Walker, J. R. (2007). Re-situating and re-mediating the canons: A cultural-historical remapping of rhetorical activity. *Kairos, 11(3)*. http://kairos.technorhetoric.net/11.3/binder.html?topoi/prior-et-al/index.html

Van Ittersum, D. (2009). Distributing memory: Rhetorical work in digital environments. *Technical Communication Quarterly, 18(3)*, 259–280.

Recommended Resources

Boer, R. (2013). Revolution in the event: The problem of kairos. *Theory, Culture, and Society, 30(2)*, 116–134. doi:10.1177/0263276412456565

Carruthers, M. (1990). *The book of memory: A study of memory in medieval culture*. Cambridge University Press.

Houdek, M., & Phillips, K. R. (2020). Rhetoric and the temporal turn: Race, gender, temporalities. *Women's Studies in Communication, 43(4)*, 369–383. https://doi.org/10.1080/07491409.2020.1824501

Reynolds, J. F. (1993). *Rhetorical memory and delivery: Classical concepts for contemporary composition and communication*. Routledge.

Whittemore, S. (2015). *Rhetorical memory: A study of technical communication and information management*. The University of Chicago Press.

See also: Diary Study, Curiosity, Reflection

27. Multigenre

Megan Marshall

Multigenre refers to a scholarly, creative, or hybrid work that incorporates a variety of genres to address a common topic or theme. Although the term is frequently linked to academic research and/or writing projects, multigenre artifacts are present across a diverse range of mediums, such as literature, hypermedia projects, and performances.

As a pedagogical strategy, multigenre is often associated with Tom Romano's (1995) work in the fields of writing and English education. His book, *Writing with Passion: Life Stories, Multiple Genres*, introduced teachers to multigenre writing as a methodological approach that deviates from teaching the traditional English research paper. His follow up, *Blending Genres, Altering Style: Writing Multigenre Papers* (Romano, 2000), framed multigenre papers as "composed of many genres and subgenres, each piece self-contained, making a point of its own, yet connected by theme or topic and sometimes by language, images and content" (x–xi). Romano includes samples that highlight how a variety of genres—such as poems, narratives, annotated bibliographies, encyclopedia entries, illustrations, and letters—emphasize how students showcase their capacity to express ideas in emotional, creative, and academic ways.

In post-secondary environments, the practice of multigenre research and writing is described as a way to provide students with agency over their scholarly work in order to foster creativity and motivation (Davis & Shadle, 2000). Additionally, it is seen as a means of potentially disrupting conventionally linear systems of reading, writing, and productions of knowledge, making space for more meaningful connections between readers, writers, and texts (Jung, 2005). To ensure that multigenre writing and research at the college level serves university expectations of rigor, Nancy Mack (2002) stresses that students should make explicit which pieces of work are grounded in fact vs. fiction. Mack hosts a website, Multigenre Report Writing, that provides a variety of exemplars from undergraduate student projects. This approach is also a natural fit for scholarly work done at the graduate level (Starke-Meyerring, Paré, & McAlpine, 2009), and a growing number of doctoral dissertations are integrating multiple genres as well, such as Anne von Petersdorff's hybrid dissertation project (2018).

Despite its habitual association with academic writing, the multi-genre approach has long made its mark in art and literature. For instance, during the latter years of the Ming dynasty in the mid-sixteenth century, Zhou Lujing published a series of painting manuals that provided painting instruction via illustrations, informational text, narrative, and cultural critique (Park, 2012). A range of authors have also made prolific use of the form, mixing crime, magical realism, journalism, sci-fi, romance, historical fiction, and paranormal genres (Danielson, 2005; Williams, 2015). A notable example is Jennifer Egan's (2011) Pulitzer Prize winning novel, *A Visit From the Goon Squad* in which chapters shifted between narrative perspective, genre (sci-fi), and/or format (text messages, a PowerPoint).

Outside the page, multigenre performances combine multiple perspectives, mixing poetry, dance, spoken word, narrative, and fine art. Hypermedia's role in writing and artistic production has also increased the scope of multigenre work, providing interactive spaces that encourage various forms of audience participation and civic discourse (Richardson, 2018).

References

Danielson, K. E., & Harrington, J. (2005). From the popcorn book to popcorn! Multigenre children's books. *Reading Horizons, 46*(1), 45–61.

Davis, R., & Shadle, M. (2000). "Building a mystery": Alternative research writing and the academic act of seeking. *College Composition and Communication, 51,* 417–446.

Egan, J. (2011) *A visit from the goon squad.* Corsair Publishers.

Jung, J. (2005). *Revisionary rhetoric, feminist pedagogy, and multigenre texts.* Southern Illinois University Press.

Mack, N. (2002). The ins, outs, and in-betweens of multigenre writing. *English Journal, 92*(2), 91–98.

Park, J. P. (2012). *Art by the book: Painting manuals and the leisure life in late Ming China.* University of Washington Press.

Richardson, J. E. (2018). Mediating national history and personal catastrophe: Televising Holocaust memorial day commemoration. *Fudan Journal of the Humanities and Social Sciences, 11.* https://doi.org/10.1007/s40647–017–0209–4

Romano, T. (2000). *Blending genre, altering style: Writing multigenre papers.* Heinemann.

Romano, T. (1995). *Writing with passion: Life stories, multiple genres.* Boynton/Cook Publishers.

Starke-Meyerring, D., Pare, A., McAlpine, L. (2009). The dissertation as a multi-genre: Many readers, many readings. In C. Bazerman, D.

Figueiredo, and A. Bonini (Eds.), *Genre in a changing world* (pp. 179–193). Parlor Press and WAC Clearinghouse.

von Petersdorff, A. (2018). Hybrid dissertation. https://www.annevon-petersdorff.com/hybrid-dissertation/

Williams, P. A. (2015). Writing the polyphonic novel. *Writing in Practice*, 1. https://www.nawe.co.uk/DB/wip-editions/articles/writing-the-poly-phonic-novel.html

Recommended Resources

Bawarshi, A. (2003). *Genre and the invention of the writer: Reconsidering the place of invention in composition.* Utah State University Press.

Bazerman, C. (1994). Systems of genres and the enactment of social intentions. In A. Freedman and P. Medway (Eds.), *Genre and the new rhetoric* (pp. 79–101). Taylor & Francis.

Giola, M. (2016). Karen Finley and Ike Holter will be part of Steppen-wolf multi-genre performance series. *Playbill.* http://www.playbill.com/article/karen-finley-and-ike-holter-among-steppenwolf-multi-genre-performance-series

Miller, C. R. (1984). Genre as social action. Quarterly Journal of Speech, 70(2), 151–167. doi:10.1080/00335638409383686

Schryer, C. F. (2018). Genre time/space: Chronotopic strategies in the experimental article. In C. R. Miller and A. J. Devitt (Eds.), *Landmark essays on rhetorical genre studies* (pp. 154–161). Routledge

See also: Multimodality, Materiality, Assemblage

28. Perseverance

Megan Poole

O ne hallmark behavior of makerspaces—perseverance—manifests in the moments following failure. When makers encounter problems in which their challenges exceed their skill level, failure is inevitable. Part of the "maker mindset" (Dougherty, 2013) is to persist in these moments of recalcitrance, to strive toward the established end goal despite setbacks, to believe that continual hard work will produce a desirable outcome. In other words, perseverance is vital throughout the making process. Not only about hard work or confidence in one's ability, however, perseverance also entails having a clear objective and maintaining the conviction that this objective matters. Put most succinctly by educational psychologist Laila Y. Sanguras (2017), perseverance is "a purposeful action to pursue a goal or task despite obstacles" (p. 11).

If making entails a recursive, iterative loop of exploration, prototype, failure, and revision, then perseverance is the spirit that moves makers back to exploration and revision as many times as is necessary to reach a desirable outcome. Psychological studies of "success" show that individuals who achieve the highest level of expertise are those who have persevered and worked the hardest, as opposed to those with "raw talent" (Duckworth, 2016). Perseverance, then, emerges as a behavior that may be built through making, as the work accomplished in makerspaces has less to do with talent, intelligence, or technical skill and more to do with the willingness to think, act, and work beyond established limits. Makers come to know through experience what Sanguras (2017) defines as a necessary tenet of education: "[Individuals] know what it feels like to want to give up—but then they persevere through it" (p. 20). Rather than giving up or professing an inability to accomplish a task, perseverance allows makers to proclaim that they may not be able to accomplish something . . . *yet*.

To ensure that perseverance remains a hallmark of makerspaces, two key guidelines should be observed at the onset of any project. First, makers should articulate their own objectives and desired outcomes. Individuals rarely persist in prescribed tasks if they do not understand the project's rationale or believe in its significance (Duckworth, 2016). Second, an environment of play should precede the articulation of outcomes and objectives. Such play promotes "small wins" that instills the belief that continual effort will lead to eventual resolution (Duckworth, 2016). Makers are always on their way to something new, and perseverance offers them the energy needed to keep persisting, improvising, making.

Fig. 28.1. Mistakes can lead to enlightenment. Photo by Megan Lee on Unsplash. https://bit.ly/keyword-fig-28–1

As makers persevere through the life cycle of a problem, from failure to a desired outcome, they come to consider the work of troubleshooting future challenges as an inevitable, if not rewarding, part of the process. Psychologist Angela Duckworth (2016) argues that the spirit of perseverance brings individuals to "interpret failure as a cue to try harder rather than as confirmation that they lack the ability to succeed" (p. 179). Instead of a qualitative judgment on the efficacy or "success" of an outcome, perseverance calls attention to what may be learned *through* the making.

References

Dougherty, D. (2013). The maker mindset. In M. Honey & D.E. Kanter (Eds.), *Design, make, play: Growing the next generation of STEM innovators* (pp. 7–11). New Routledge.

Duckworth, A. (2016). *Grit: The power of passion and perseverance.* Scribner.

Sanguras, L. Y. (2017). *Grit in the classroom: Building perseverance for excellence in today's students.* Prufrock Press.

Recommended Resources

Hughes, J. M. (2017). Digital making with "at-risk" youth. *The International Journal of Information and Learning Technology, 34*(2), 102–113.

Laursen, E. K. (2015). The power of grit, perseverance, and tenacity. *Reclaiming Children and Youth, 23*(4), 19–24.

Steier, L., & Young, A. W. (2016). *Growth mindset and the makerspace educational environment.* Sophia, St. Catherine University repository. https://sophia.stkate.edu/maed/196/

Tham, J. (2021). *Design thinking in technical communication: Solving problems through making and collaboration.* Routledge.

Weng, Z., Chiu, T. K. F., & Tsang, C. C. (2022). Promoting student creativity and entrepreneurship through real-world problem-based maker education. *Thinking Skills and Creativity, 45,* 101046. https://doi.org/10.1016/j.tsc.2022.101046

See also: Creative Confidence, Bias Toward Action, Safety

29. Play/Playful

A. Nicole Pfannenstiel

Play is often conflated with games and children; however, play is a complex social action, performed by children and adults alike, that includes making and critical thinking (Huizinga, 2006). In *The Maker Movement Manifesto*, Mark Hatch (2014) lists play as a key component to making, describing laughter and experimentation as what play offers making (pp. 26–28). As most game players can attest, playing is not always fun, but often tedious experimentation and risk taking.

Fig. 29.1. Play as an integral part of life. Photo by zhang kaiyv on Unsplash. https://bit.ly/keyword-fig-29–1

At the core of play (the activity) is a playful habit of mind, an attitude or mindset individuals bring to situations and spaces (Stenros, 2014). Playfulness is creativity and imagination, a way of engaging the world (Resnick, 2017). Additionally, playful approaches can demonstrate how critical thinking is developed and reinforced through play (Resnick, p. 130). Playfulness is also liberating; it aids the player in considering new ways of approaching tools and artifacts to break beyond expected uses, to (re)make. Playfulness is about tinkering with ideas, uses, tools, materials, and spaces in creative and critical ways. The inherent pleasure of play comes from creatively remixing across structured domains/elements. Mary Flanagan (2013) extends this further, arguing that "critical play is characterized by a careful examination of social, cultural, political, or even personal themes that function as alternatives to popular play spaces" (p. 6). Makers approaching tools and artifacts with a critically playful mindset employ that mindset as a framework for considering and critiquing social expectations of the ideas, tools, materials and spaces to (re)imagine ways of making, designing, and doing that question inherent assumptions about use, materials, ideas, spaces, and society.

Playfulness is located within an individual (Stenros, 2014) who must be willing to play in a space. For instance, moving from fun to critical play requires the maker to engage playfully, which can lead to subversion through play. The coding robots *Ozobots* (ozobot.com) are a great case study for playfulness. These small "robots" teach makers to code using combinations of marker colors used to make shapes and designs the robot "reads" (portal.ozobot.com/lessons). This design can be playfully subverted as makers imagine how to exploit the programming. Instead of just following the designed codes, makers can represent ideas, stories, narratives, writing processes and more through a series of lines and shapes. Makers can design using materials and tools from a variety of sources, then use marker lines to help the Ozobot navigate. This playful subversion asks makers to think through how to represent in ways that can be traversed by the robot. When makers playfully engage, this activity can be creative and fun, but also emancipatory and liberating. While the robots are designed to teach coding, this subversion of designed use creates space for makers to bridge social, cultural, even political spaces in their playful making. The robot traverses; the maker plays.

Play is never *just play* but an activity of possibilities within spaces. Playfully approaching doing and making allows players to not only engage tools, materials, and spaces to make and do, but to critically engage tools, materials, and spaces to make and do in ways that can subvert inherent assumptions (re)imagining possibilities. A playful mindset empowers makers.

References

Flanagan, M. (2013). *Critical play: radical game design*. MIT Press.

Hatch, M. (2014). *The maker movement manifesto: rules for innovation in the new world of crafters, hackers, and tinkerers*. McGraw-Hill Education.

Huizinga, J. (2006). Nature and significance of play as a cultural phenomenon. In K. Salen and E. Zimmerman (Eds.), *The game design reader: a rules of play anthology* (pp. 96–119). MIT Press.

Resnick, M. (2017). *Lifelong kindergarten: cultivating creativity through projects, passion, peers, and play.* MIT Press.

Stenros, J. (2014). Behind Games: Playful mindsets and transformative practices. In S.P. Walz and S. Deterding (Eds.), *The gameful world: Approaches, issues, applications* (pp. 201–222). MIT Press.

Recommended Resources

Han, S. Y., Yoo, J., Zo, H., & Ciganek, A. P. (2017). Understanding makerspace continuance: A self-determination perspective. *Telematics and Informatics, 34*(4), 184–195.

Horst, R., James, K., Takeda, Y., Rowluck, W. (2020). From play to creative extrapolation: Fostering emergent computational thinking in the makerspace. *Sustainability, 15*(5), 40–54. https://www.proquest.com/scholarly-journals/play-creative-extrapolation-fostering-emergent/docview/2472178880/se-2

Pettersen, I. B., Kubberød, E., Vangsal, F., & Zeiner, A. (2020). From making gadgets to making talents: Exploring a university makerspace. *Education + Training, 62*(2), 145–158. https://www.emerald.com/insight/content/doi/10.1108/ET-04-2019-0090/full/html

Sutton-Smith, B. (1997). *The ambiguity of play.* Harvard University Press.

Voussoghi, S. and Bevan, B. (2014). Making and tinkering: A review of the literature. National Research Council Committee on Out of School STEM, National Research Council, 1–55. http://sites.nationalacademies.org/cs/groups/dbassesite/documents/webpage/dbasse_089888.pdf

See also: Curiosity, Creative Confidence, Tinkering, Hacking

30. Postdigital Aesthetics

Jialei Jiang

In some ways, the maker movement and its culture reside in what scholars identify as postdigital aesthetics. Postdigital aesthetics is rooted in the contemporary disenchantment with the "obsessively digital" (Portanova, 2015) components in a wide array of fields, including music, architecture, design, and media studies, among other disciplines. In 2013, a meme went viral that depicted a man using a typewriter rather than a computing device while he was sitting on a park bench. The caption of the meme reads, amusingly and sarcastically: "You're not a real hipster—until you take your typewriter to the park" (Cramer, 2015, p. 12). The meme encapsulates the tension between digital and post-digital cultures (Cramer, 2015). While far from indicating a rejection of the digital, postdigital aesthetics is a movement toward a more comprehensive approach to the digital beyond simply equating technology with improvement and advancement in society (Cramer, 2015). In troubling techno-optimism, the postdigital turn (Pepperell & Punt, 2000; Taffel, 2016) prompts scholars to recognize the complex consequences of technology for issues related to digital media, literacy, and information ecologies.

Postdigital aesthetics (Bridle, 2010; Berry & Dieter, 2015; Cramer, 2015; Pepperell & Punt, 2000) takes issue with the separation between the digital and the analogue, technology and human in media studies scholarship. Moving beyond the binary logic in existing ideologies (Pepperell & Punt, 2000), the postdigital turn calls our attention to the embodiment of technologies in our daily life and the material impact of the digital in human-machine relationships. Sy Taffel (2016) similarly urges scholars to combat the dualistic logic grounded in the view of the digital as equivalent to "the discrete samples of binary codes" (p. 326). Rather, Taffel (2016) forwards an understanding of the (post)digital through the lens of hypermediation that involves considering technology as embedded in technocultural assemblages. In other words, postdigital signals a rejection of oppositions to support a holistic view of the complex networks and materiality—including both human and nonhuman actors—that technologies are dependent on.

Since its inception, the concept of postdigital aesthetics has been taken up in diverse interdisciplinary conversations across new media studies (Bevan, 2018; Taffel, 2016; Thorén et al., 2019), game studies (Apperley et al., 2016; Aarseth, 2006; Holloway-Attaway & Rouse, 2018; Jayemanne et al, 2016), as well as science and education research (Bhatt & MacKenzie, 2019; Jandri et al., 2018; Jiang & Vetter, 2020). In comparison, the foray of postdigital aesthetics into rhetorical scholarship and related fields has only been a recent phenomenon. One example is Justin Hodgson's (2019) *Post-Digital Rhetoric and the New Aesthetic,* in which Hodgson suggests that digital rhetoricians recover the visual and aesthetic dimensions of knowledge that have long been absent in the tradition of digital rhetoric. Through taking an aesthetic perspective on rhetorical practices, Hodgson focuses his discussion on four interrelated conceptions of postdigital rhetoric, including 1) framing a rhetorical ecology in recognizing human-technology assemblages, 2) highlighting the capacity of technology that in (re)shaping human sensibilities, 3) detailing the collaboration between humans and technologies in making rhetoric, and 4) mapping the value of hypermediation in mobilizing rhetorical practices. Taken together, Hodgson (2019) deploys postdigital aesthetics to transform our current understanding and practice of digital rhetoric.

References

Aarseth, E. (2006). How we become postdigital. In D. Silver & A. Massanari (Eds.), *Critical cyberculture studies* (pp. 37–46). New York University Press.

Apperley, T., Jayemanne, D., & Nansen, B. (2016). Postdigital literacies: Materiality, mobility and the aesthetics of recruitment. In B. Parry, C. Burnett, and G. Merchant (Eds.), *Literacy, media, technology: Past, present and future* (pp. 203–218). Bloomsbury Publishing.

Bevan, A. (2018). 3-D printed sets and props: How designers integrate digital technologies. *Convergence, 24*(6), 554–567.

Berry, D. M., & Dieter, M. (2015). Thinking postdigital aesthetics: art, computation and design. In D. M. Berry and M. Dieter (Eds.), *Postdigital aesthetics* (pp. 1–11). Palgrave Macmillan.

Bhatt, I., & MacKenzie, A. (2019). Just Google it! Digital literacy and the epistemology of ignorance. *Teaching in Higher Education, 24*(3), 302–317.

Bridle, J. (2010). Web Directions South 2010 Interview. http://www.webdirections.org/blog/ james-bridle-web-directions-south-2010-interview/

Cramer, F. (2015). What is "Post-digital"? In D.M. Berry and M. Dieter (Eds.), *Postdigital aesthetics* (pp. 12–26). Palgrave Macmillan.

Hodgson, J. (2019). *Post-digital rhetoric and the new aesthetic*. The Ohio State University Press.

Holloway-Attaway, L., & Rouse, R. (2018). Designing postdigital curators: establishing an interdisciplinary games and mixed reality cultural heritage network. In M. Ioannides, J. Martins, R. Žarnić, and V. Lim (Eds.), *Advances in digital cultural heritage* (pp. 162–173). Springer, Cham.

Jandrić, P., Knox, J., Besley, T., Ryberg, T., Suoranta, J., & Hayes, S. (2018). Postdigital science and education. *Educational Philosophy and Theory, 50*(10), 893–899.

Jiang, J., & Vetter, M. (2020). The good, the bot, and the ugly: Problematic information and critical media literacy in the postdigital era. *Postdigital Science and Education, 2*(1), 78–94. https://link.springer.com/article/10.1007/s42438-019-00069-4

Pepperell, R., & Punt, M. (2000). *The postdigital membrane: Imagination, technology and desire.* Intellect Books.

Portanova, S. (2015). The genius and the algorithm: Reflections on the new aesthetic as a computer's vision. In D.M. Berry and M. Dieter (Eds.), *Postdigital aesthetics* (pp. 96–108). Palgrave Macmillan.

Taffel, S. (2016). Perspectives on the postdigital: Beyond rhetorics of progress and novelty. *Convergence, 22*(3), 324–338.

Recommended Resources

Andrews, I. (2013). Post-digital aesthetics and the function of process. In K. Cleland, L. Fisher, and R. Harley (Eds.), *Proceedings of the 19th International Symposium on Electronic Art, ISEA2013 Sydney* (pp. 1–3). http://hdl.handle.net/2123/9688

Betancourt, M. (2017). Glitch art in theory and practice: Critical failures and post-digital aesthetics. Routledge Focus.

Menkman, R. (2011). *The glitch moment(um).* Colophon.

Morris, R. (1970). Some notes on the phenomenology of making. *Artforum, 8*(8), 62–66.

Thorén, C., Edenius, M., Lundström, J. E., & Kitzmann, A. (2019). The hipster's dilemma: What is analogue or digital in the post-digital society? *Convergence, 25*(2), 324–339.

See also: Posthuman Practice, Digital Rhetoric, Interactive Design

31. Posthuman Practice

Jialei Jiang

Emerging from philosophical, scientific, and technocultural conversations, posthuman theories (Braidotti, 2013; Hayles, 2008; Haraway, 1991; Ferrando, 2013; Pepperell, 1995; Wolfe, 2010) warn against the dualistic thinking looming behind mind/body, human/nonhuman binaries, through recasting agency as dispersed across human and machine, material and information boundaries. Design can be considered a complex exercise caught in the discussions of posthuman practice. One case in point is the feminist metaphor of cyborg, or "a cybernetic organism, a hybrid of machine and organism, a creature of social reality as well as a creature of fiction" (Haraway, 1991, p. 149). In digital media spaces, human-technology interrelation exercises its agential effects through taking the shape of a hybrid creature.

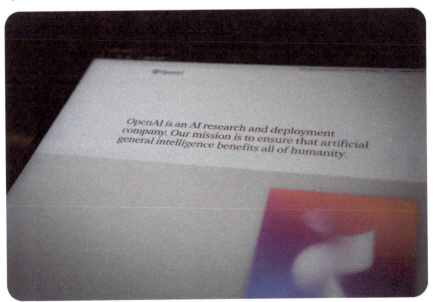

Fig. 31.1. Generative artificial intelligence programs such as OpenAI (http://openai.com) took the world by storm in 2023 after it released its free conversational AI writer. Photo by Jonathan Kemper on Unsplash. https://bit.ly/keyword-fig-31–1

Agency has been reframed as embodied, embedded, and enacted in lived experiences and flexible relations that do not always result solely from human consciousness, from either individual intentions or social discourses. The posthuman movement in rhetoric and composition (Barnett & Boyle, 2016; Boyle, 2016, 2018; Hallenback, 2012; Mays, Rivers, & Sharp-Hoskins, 2017; Muckelbauer & Hawhee, 2000), similarly signals a shift away from the human-centric view of rhetorical practices and toward articulating the various networks of human, biological, and technological agents. For instance, through case studies of the art of glitch, social media use, and DIY digital networks, Casey Boyle's (2018) *Rhetoric as a Posthuman Practice* sheds light on the embodied practice of information by delineating the key role played by various media in mediating and organizing emerging bodies.

In response to the posthuman turn, scholars in rhetorical research, design studies, and technical communication (Garrison, 2018; Kimbell, 2012; Lupton, 2018; Mara & Hawk, 2010; Moore & Richards, 2018) have sought to explore the implications of a posthuman theory for communication in technologically-mediated classrooms and workplaces. In their introduction to the *Technical Communication Quarterly* special issue titled "Posthuman Rhetoric and Technical Communication," Andrew Mara and Byron Hawk (2010) reconstrue professional and technical writing as being and having always been posthuman practice. They make clear that posthumanism holds relevance for addressing the complexity of rhetorical situations by attending to the "complex interplays among human intentions, organizational discourses, biological trajectories, and technological possibilities" (p. 3). As digital technologies become increasingly ubiquitous, it is no longer feasible to separate human action from technologies and environments (Mara & Hawk, 2010). Contributors to the special issue advance our understanding of posthuman practices through exploring the connection between work in the field and conversations surrounding posthumanism.

Building off Mara and Hawk's work, Kristen R. Moore and Daniel Richards (2018) have co-edited a collection that centers on exploring posthuman praxis and its value in initiating social changes. Posthuman praxis broadens the scope of research and practice to further explore the social, historical, political, ecological implications of posthuman theory for technical and professional communication (Moore & Richards, 2018). Contributors share both strengths and concerns over the deployment of posthumanism. For instance, in discussing its application in social justice research, Emma J. Rose and Rebecca Walton (2018), while valorizing the benefits of posthumanism in effecting changes, caution that a focus on ascribing equality to nonhuman actors may distract us from centering on social injustice experienced by marginalized people. All together, the collection envisions the potential of posthuman perspectives to transform technical communication practices.

References

Barnett, S., & Boyle, C. A. (Eds.). (2016). *Rhetoric, through everyday things*. University of Alabama Press.

Braidotti, R. (2013). *The posthuman*. John Wiley & Sons.

Boyle, C. A. (2016). Writing and rhetoric and/as posthuman practice. *College English, 78*(6), 532–554.

Boyle, C. A. (2018). *Rhetoric as a posthuman practice*. The Ohio State University Press.

Ferrando, F. (2013). Posthumanism, transhumanism, antihumanism, metahumanism, and new materialisms. *Existenz, 8*(2), 26–32.

Garrison, K. (2018). Moving technical communication off the grid. *Technical Communication Quarterly, 27*(3), 201–216.

Hallenbeck, S. (2012). Toward a posthuman perspective: Feminist rhetorical methodologies and everyday practices. *Advances in the History of Rhetoric, 15*(1), 9–27.

Haraway, D. J. (1990). A manifesto for cyborgs: Science, technology, and socialist feminism in the 1980s. In D.J. Haraway, *Simians, cyborgs, and women: The reinvention of nature* (pp. 65–107). Routledge.

Hayles, N. K. (2008). *How we became posthuman: Virtual bodies in cybernetics, literature, and informatics*. University of Chicago Press.

Kimbell, L. (2012). Rethinking design thinking: Part II. *Design and Culture, 4*(2), 129–148.

Lupton, D. (2018). Towards design sociology. *Sociology Compass, 12*(1), n.p.

Mara, A., & Hawk, B. (2010). Posthuman rhetorics and technical communication. *Technical Communication Quarterly, 9*(1), 1–10.

Mays, C., Rivers, N. A., & Sharp-Hoskins, K. (Eds.). (2017). *Kenneth Burke + the posthuman* (Vol. 6). Penn State Press.

Moore, K. R., & Richards, D. P. (Eds.). (2018). *Posthuman praxis in technical communication*. Routledge.

Muckelbauer, J., & Hawhee, D. (2000). Posthuman rhetorics: "It's the future, Pikul." *JAC*, 767–774.

Pepperell, R. (1995). *The post-human condition*. Intellect books.

Rose, E. J., & Walton, R. (2018). Factors to actors: Implications of posthumanism for social justice work. In K.R. Moore & D.P. Richards (Eds.), *Posthuman praxis in technical communication* (pp. 91–117). Routledge.

Wolfe, C. (2010). *What is posthumanism?* University of Minnesota Press.

Recommended Resources

Braidotti, R. (2017). Four theses on posthuman feminism. In R. Grusin (Ed.), *Anthropocene feminism* (pp. 21–48). University of Minnesota Press.

Dinerstein, J. (2006). Technology and its discontent: On the verge of the posthuman. *American Quarterly, 58*(3), 569–595. https://www.jstor.org/stable/4006838 4

Herbrechter, S. (2013). *Posthumanism: A critical analysis*. Bloomsbury.

Roden, D. (2015). *Posthuman life: Philosophy at the edge of the human*. Routledge.

Simon, B. (2003). Introduction: Toward a critique of posthuman futures. *Cultural Critique, 53*, 1–9. https://muse.jhu.edu/article/41118

See also: Postdigital Aesthetics, Materiality, Embodiment

32. Radical Imagination

Stephanie West-Puckett

Transnational capitalism—marked by growing injustice and inequity coupled with a shrinking public commons—has produced deleterious effects on individuals, communities, and our collective sense of agency over our own lives. As Henry Giroux (2007) illustrates, the problems of injustice and inequity created by capitalism seem too big, too diffuse, and too far gone. In this era of excessive suffering and increasing isolation from the public sphere, we've resigned ourselves to ambivalence (at best) and cynicism (at worst), and we've accepted our current situation as subjects incapable of intervening in the politics of our everyday lives.

Social theorists understand this state of impotence as the condition that ensues when we fail to activate the *radical imagination*. According to Max Haiven & Alex Khasnabish (2010) the radical imagination enables us "to imagine the world, social institutions and human (and non-human) relationships *otherwise . . .*" [emphasis added]. It is not a singular or monolithic construct; instead, the radical imagination is a method of social activation and activism that generates a multiplicity of iterative possibilities for the future—"ways of living otherwise, of cooperating differently, that reject, strain against, or seek to escape from the capitalist, racist, patriarchal, heteronormative, colonial, imperial, militaristic, and fundamentalist forms of oppression that undergird our lives" (Haiven & Khasnabish, 2010, p. xxviii). Gary A. Olson and Lynn Worsham with Giroux (2007) argue that the radical imagination is enabled through a politics of hope. Their conceptualization of hope rejects a naive, spiritual or metaphysical construction of hope. Instead, they advocate for the critical appraisal of power relations and material conditions in the present, robust questioning through public discourse and dialogue, and collective imaginings that can restructure relations and conditions to end suffering and oppression.

Both a product of and a material structure to sustain the radical imagination, makerspaces can promote sharing economies that refuse the framework of capitalism for engagement. They sponsor creativity, enable alternate forms of cultural production, and can empower counter-hegemonic cultural relations. Animated by entanglements of people, objects, technologies, and shared ideologies, makerspaces give form and structure to the do-it-yourself (DIY) movement whose participants share disillusionment with the vacuity of mainstream, globalized cultural production. Makerspaces bring people with shared DIY affinities together and foster the collective conditions necessary for the emergence of do-it-together (DIT) or do-it-with-others (DIWO) ethic. Through the solidarity of DIT/DIWO, makers engage in the creation of art, politics, and technological solutions, "for reasons other than profit" (Griffin, 2015, p. 124) and practice anti-normative ways of knowing, doing, and being together in the world.

Examples of makerspaces that sustain and nurture the radical imagination include Incite Focus (https://www.incite-focus.org/) and Xerocraft Hackerspace (https://www.xerocraft.org/about.php). According to Will Holman (2015), Incite Focus is exploring "the kind of community-based production that can support a sustainable local economy . . . using proven old technologies (like permaculture) and updating them with agile new technologies (like digital fabrication)." Similarly, at Xerocraft Hackerspace, the radical imagination is fostering interventions such as their *WTF! Women, Trans, and Femme* (WTF) night. One night a week, for a four-hour block, the space is facilitated by and reserved for WTF-identified makers, providing opportunities to redistribute technical and vocational composing expertise among those who have historically been marginalized.

While these two examples showcase makers working collectively to compose new selves, new communities, and new relations as a way of ushering in more democratic futures, it is important to note that not all makerspaces are positioned to remake the social, political, and economic conditions of late modernity. In fact, many makerspaces, particularly those that function as for-profit entities, intentionally promote a boot-strapping approach that is designed to support individuals in acquiring marketable technology skills which will help them enter, succeed, and profit within the structures of capitalism. Makerspaces that sponsor market-driven technological literacies are complicit in promoting both individualism and competition and privilege male literacies and financial independence as they prepare makers to take advantage of the next potentially profitable venture. These spaces remind us that without the radical imagination, the maker movement can be easily subsumed by entrepreneurialism, neoliberalism, and the all-encompassing logics of commodification.

References

Giroux, H. (2007). Foreword. In G. Olson and L. Worsham (Eds.), *Politics of possibility: Encountering the radical imagination* (pp. vii–xvii). Taylor and Francis.

Griffin, N. (2015). *Understanding DIY punk as activism: Realizing DIY ethics through cultural production, community and everyday negotiations.* Doctoral dissertation. Northumbria University. http://nrl.northumbria.ac.uk/30251/

Haiven, M., & Khasnabish, A. (2010). What is radical imagination? A special issue. *Affinities: A Journal of Radical Theory, Culture, and Action* 4(2), i–xxxvii.

Holman, W. (2015). Makerspace: Towards a new civic infrastructure. *Places Journal.* https://doi.org/10.22269/151130

Recommended Resources

Ginwright, S. (2008). Collective radical imagination: Youth participatory action research and the art of emancipatory knowledge. In J. Cammarota and M. Fine (Eds.), *Revolutionizing education: Youth participatory action research in motion* (pp. 13–22). Routledge.

Green, K. L. (2020). Radical imagination and "otherwise possibilities" in qualitative research. *International Journal of Qualitative Studies in Education,* 33(1), 115–127. https://doi.org/10.1080/09518398.2019.1678781

Haiven, M., & Khasnabish, A. (2014). *The radical imagination: Social movement research in the age of austerity.* Fernwood Publishing.

Lewis, D. (2007). Feminism and the radical imagination. *Agenda, 21*(72), 18–31. https://doi.org/10.1080/10130950.2007.9674832

Vrettos, J. (n.d.). *The radical imagination.* Manhattan Neighborhood Network. YouTube (Playlist). https://www.youtube.com/playlist?list=PLF6GAPebezXm9MUWZBn_jMXyE6JIb3dO7

See also: Invention, Shared Leadership, Radical Collaboration

33. Social Constructivism

Laura Roberts

Designing and making involve considering the multiple possibilities of an artifact and limiting those possibilities into a defined object through rhetorical and technical choices. Social constructivism works backwards to interrogate the multiple possibilities and trajectories inherent in any designed artifact. Social constructivism is a methodology arising out of sociology programs beginning in the 1970s and early 1980s that was largely influential to humanities studies, particularly the study of Science, Technology, and Society (STS) in the 1980s and 1990s (Lynch, 2016). Researchers who deploy social constructivism examine the historical, cultural, and social origins of theories, facts, tools, and technologies typically through case studies.

There are two prominent and interrelated branches of social constructivism research: studies of scientific knowledge and social construction of technology (SCOT). For science studies scholars, social constructivism frequently involves "opening the black box" of science to reveal the social processes and negotiations by which scientific knowledge becomes accepted as a theory or fact (Winner, 1993). Representative of this work is Bruno Latour and Steve Woolgar's (1979) *Laboratory Life: The Construction of Scientific Facts* in which Latour and Woolgar treat the Salk laboratories as an anthropological site. Indicative of social constructivist work, Latour and Woolgar do not take scientific facts as "natural" or "given," but instead, pursue the historicity of statements which become more or less "fact-like" (Latour & Woolgar, 1979).

Technology studies scholars similarly deploy social constructivism to study the development of technologies and technological artifacts. Their use of social constructivism is typically associated with the social construction of technology, or SCOT. A key concept of SCOT is interpretative flexibility—a stance toward technology that acknowledges its contingency. Similar to questioning the naturalness of scientific fact, interpretative flexibility questions the innateness of technology in its design, application, and social effects. For example, a hammer could be designed in any myriad of ways. The design of the hammer arises from consensus among relevant social groups over time who determine what makes a hammer a hammer. A well-known example of social constructivism is Trevor Pinch and Wiebe Bijker's (1987) description of the development of the safety bike. Although multiple iterations of the bicycle were developed that catered to different social groups, eventually one form of the bicycle was embraced over others due to social negotiations and consensus.

Fig. 33.1. Technology mediates how we interact with our social environment and with one another. Photo by Eaters Collective on Unsplash. https://bit.ly/keyword-fig-33–1

Social constructivism is often contrasted with technological determinism, in which technological determinists argue that technologies have certain political and social consequences in the world. While a technological determinist would focus on the political organization that a technology entails, such as how nuclear technology requires centralized forms of government (Winner, 1980), a social constructivist would contemplate the social negotiations that enabled nuclear technology to establish.

While sometimes critiqued for its epistemic relativism (Winner, 1993), social constructivism is praised for its ability to produce in-depth and comprehensive accounts of scientific and technical knowledge, such as Donald Mackenzie and Graham Spinardi's (1995) sociological history of nuclear weapons technology. Social constructivism continues to be highly influential to STS studies as well as the fields of media studies (Fischer, 1994) and writing studies (Callon, 2002). A separate, but related, form of social constructivism is employed in educational psychology, deriving from the works of Vygotsky (Hodson & Hodson, 1998).

References

Callon, M. (2002). Writing and (re)writing devices as tools for managing complexity. *Complexities: Social Studies of Knowledge Practices*, 191–217.

Fischer, C. S. (1994). *America calling: A social history of the telephone to 1940*. University of California Press.

Hodson, D., & Hodson, J. (1998). From constructivism to social constructivism: A Vygotskian perspective on teaching and learning science. *School Science Review*, 79(289), 33–41.

Latour, B., & Woolgar, S. (1979). *Laboratory life: The construction of scientific facts*. Sage Publications.

Mackenzie, D., & Spinardi, G. (1995). Tacit knowledge, weapons design, and the uninvention of nuclear weapons. *American Journal of Sociology*, 101(1), 44–99.

Pinch, T. J., & Bijker, W. E. (1984). The social construction of facts and artefacts: Or how the sociology of science and the sociology of technology might benefit each other. *Social Studies of Science*, 14(3), 399–441.

Lynch, M. (2016). Social constructivism in science and technology studies. *Human Studies*, 39(1), 101–112.

Winner, L. (1993). Upon opening the black box and finding it empty: Social constructivism and the philosophy of technology. *Science, Technology, & Human Values*, 18(3), 362–378.

Winner, L. (1980). Do artifacts have politics?. *Daedalus*, 121–136.

Recommended Resources

Bijker, W. E., Hughes, T. P., & Pinch, T. (Eds.). (1987). *The social construction of technological systems: New directions in the sociology and history of technology*. MIT Press.

Giotta, G. (2018). Teaching technological determinism and social construction of technology using everyday objects. *Communication Teacher*, 32(3), 136–140. https://doi.org/10.1080/17404622.2017.1372589

Hacking, I. (1999). *The social construction of what?* Harvard University Press.

Klein, H. K., & Kleinman, D. L. (2002). The social construction of technology: Structural considerations. *Science, Technology, & Human Values*, 27(1), 28–52.

Olsen, O. E., & Engen, O. A. (2007). Technological change as a trade-off between social construction and technological paradigms. *Technology in Society*, 29(4), 456–468. https://doi.org/10.1016/j.techsoc.2007.08.006

See also: Constructionism, Active Learning, Inter/Cross-Disciplinarity

34. Visual Rhetoric

Jason Tham

Many scholars have attempted to bridge rhetoric and design. According to Victor J. Gallagher, Kelly Norris Martin, and Magdy Ma (2011), rhetoric and design are "two distinct fields of study intricately related as reflected in their assumptions, goals and function" (p. 27). The relationship between design and rhetoric is evident in that both disciplines are rooted in cultural, economic and technological developments. Designers translate concepts and ideas into a visual representation, by organizing and connecting elements into a structure. This arrangement of elements is done with an intended effect in mind—a goal. Since the communication between designers and viewers has defined purposes, design is essentially rhetorical. This intentional and deliberate production of meaning is the rhetorical function of visual design (Emanuel, 2010).

Fig. 34.1. Design elements like colors can be used to represent and augment arguments. Photo by Raphael Renter on Unsplash. https://bit.ly/keyword-fig-34–1

Gunther Kress and Theo van Leeuwen have contended that "language and visual communication can both be used to realize the 'same' fundamental systems of meaning that constitute our culture, but each does so by its own specific forms, does so differently, and independently" (2006, p. 19). Amid this claim, there seem to be conceivable commonalities between rhetorics and aesthetics, that there might be a way for theorizing the verbal and the visual using similar vocabularies or concepts. In fact, Kress and van Leeuwen were not the first to make such an assumption; German designer Gui Bonsiepe (1961, 1965) had attempted to develop explicit transfer of the language of rhetoric to the visual dating close to fifty years ago. Against a rhetorical background, Bonsiepe has developed the first analogies for understanding visual design and rhetoric, sharpened design analysis vocabularies, and showed that designers use certain defined figurations in design to enable effective communication.

Although a natural affinity exists between rhetoric and visual design, the inclusion of visual imagery in rhetorical study has not been the seamless process many might assume. Proposals to expand rhetoric to encompass the visual were at first met with objections from within its field. Waldo Braden (1970), for example, has suggested that rhetoricians are not trained to deal with visual images. Another reason cited for the reluctance of rhetoric scholars to tackle the study of visual images has had less to do with personal competencies, but rather their desire to accumulate theoretical insights into rhetoric. Roderick Hart (1976) has said that,

> To the extent that scholars deviate from traditional, commonly shared understandings of what rhetoric is—by including non-social, mechanically mediated, and nonverbal phenomena in the rhetorical mix—they are, to that extent, necessarily forsaking the immediate implementation of the theoretical threads derived in previous studies of human, non-mediated, problematic, verbal interchanges. (pp. 71–72)

Nonetheless, as we observe today, the study of visual images has continued and now flourishes in rhetorical studies thanks to the pervasiveness of the visual and its impact on many aspects of contemporary culture (Foss, 2005). The study of visual imagery from a rhetorical perspective also has grown with the recognition that the visual provides access to a range of human experience not always available through the study of verbal discourse.

Visual rhetoric emerges as a framework to understand and articulate such experiences. According to Sonja Foss (2005), visual rhetoric is used to mean both a visual object or artifact and a perspective on the study of visual data. Conceptualized as a communicative artifact, visual rhetoric is the actual visual that rhetors create for the purpose of communication—a painting, an advertisement, or a chart that constitutes the data of study. While this aspect of visual rhetoric in its broadest term mirrors what is known as design, Foss argues that three characteristics are needed to turn the visual artifact into a communicative artifact. Visual rhetoric must be symbolic, involve human intervention, and be presented to an audience in order to communicate with them (Foss, 2005, p. 144). This expands design beyond its aesthetic value to a utilitarian, purposive construction that serves particular communicative needs.

As a perspective, visual rhetoric constitutes a theoretical viewpoint that involves the analysis of the symbolic or communicative aspects of visual rhetoric. It is a critical-analytical tool for visual data that highlights the communicative dimensions of images. The key to a rhetorical perspective on images is its focus on a rhetorical response rather than just aesthetic one. For instance, colors, lines, textures, and rhythms in an image would provide a basis for the viewer to infer the existence of the image, emotions, and ideas. A visual-rhetorical perspective focuses on understanding such responses to images.

In short, visual rhetoric combines the artifact (design) and the way of viewing (rhetoric). Together, these senses of the term point to new ways of understanding how the visual operates rhetorically in contemporary culture. Visual rhetoric suggests the need to expand understanding of the multifaceted ways in which symbols—verbal as well as visual—inform and define human experiences.

References

Bonsiepe, G. (1961). Persuasive communication: Towards a visual rhetoric. In T. Crosby (Ed.), *Uppercase 5*, 19–34.

Bonsiepe, G. (1965). Visuell/verbale rhetorik (Visual/verbal rhetoric). *Ulm, 14/15/16*, 22–40.

Braden, W. (1970). Rhetorical criticism: Prognoses for the Seventies—A symposium: A prognosis by Waldo W. Braden. *Southern Speech Journal*, *36*, 104–107.

Emanuel, B. (2010). Rhetoric in graphic design. Master's thesis. Hochschule Anhalt. http://graphicdesignrhetoric.tumblr.com

Foss, S. K. (2005). Theory of visual rhetoric. In K. Smith, S. Moriarty, G. Barbatsis, and K. Kenney (Eds.), *Handbook of visual communication: Theory, methods, and media* (pp. 141–152). Lawrence Erlbaum.

Gallagher, V. J., Martin, K. N., & Ma, M., (2011). Visual wellbeing: Intersections of rhetorical theory and design. *Design Issues, 27*(2), 27–40.

Hart, R. (1976). Forum: Theory-building and rhetorical criticism: An informal statement of opinion. *Central States Speech Journal, 27*, 70–77.

Kress, G., & van Leeuwen, T. (2006). *Reading images: The grammar of visual design (2nd Edition)*. Routledge.

Recommended Resources

Buchanan, R. (1985). Declaration by design: Rhetoric, argument, and demonstration in design practice. *Design Issues, 17*(3), 3–23.

Buchanan, R. (1990). Myth and maturity: Toward a new order in the decade of design. *Design Issues, 6*(2), 70–80.

Buchanan, R. (1995). Rhetoric, humanism, and design. In R. Buchanan and V. Margolin (Eds.), *Discovering design: explorations in design studies* (pp. 23–66). University of Chicago Press.

Buchanan, R. (2001). Design and the new rhetoric: Productive arts in the philosophy of culture. *Philosophy and Rhetoric, 34*(3), 183–206.

Buchanan, R. (2007). Strategies of design research: Productive science and rhetorical inquiry. In R. Michel (Ed.), *Design research now: essays and selected projects* (pp. 55–66). Birkhäuser.

Cyphert, D. (2004). The problem of PowerPoint: Visual aid or visual rhetoric? *Business Communication Quarterly, 67*(1), 80–84.

Fried, J. (2008). Why we disagree with Don Norman. Signal v. Noise. https://signalvnoise.com/posts/904-why-we-disagree-with-don-norman

Garber, L., & Hyatt, E. (2003). Color as a tool for visual persuasion. In L.M. Scott and R. Batra (Eds.), *Persuasive imagery a consumer response perspective* (pp. 313–336). Lawrence Erlbaum.

Kaufer, D. S., & Butler, B. S. (1996). *Rhetoric and the arts of design*. Lawrence Erlbaum.

Kennedy, G.A. (1991). *Aristotle, on rhetoric: A theory of civic discourse*. Oxford University Press.

Marcus, A., Guttman, E., & Atwood, M. (1999). Visual design for E-commerce and performance tools. *ACM Human Factors in Computing Systems: CHI '99 Extended Abstracts* (pp. 112–113). ACM.

McKeon, R. (1971). The uses of rhetoric in a technological age: Architectonic productive arts. In L.F. Bitzer and E. Black (Eds.), *The prospect of rhetoric* (pp. 44–63). Prentice-Hall.

Norman, D. (2014). State of design: How design education must change. *LinkedIn Pulse*. https://www.linkedin.com/pulse/20140325102438-12181762-state-of-design-how-design-education-must-change

Sheridan, D. (2010). Fabricating consent: Three-dimensional objects as rhetorical compositions. *Computers and Composition, 27*(4), 249–265.

Silva Rhetoricae (2015). What is rhetoric? http://rhetoric.byu.edu/En-
   compassing%20Terms/rhetoric.htm

See also: Visual Semiotics, Aesthetics, Data Visualization

35. Visual Semiotics

Jason Tham

Much of the current writing studies scholarship on multimodality stems from literature on the visual mode of scientific and technical communication. For example, Charles Bazerman (1981), in his analysis of James Watson and Francis Crick's landmark article on the structure of DNA, notes their use of a diagram on their first page in order to provide "the geometrical essence of the solution" (p. 368). In their book on visual design, Gunther Kress and Theo van Leeuwen (2006) point out that visuals play a prominent role in scientific meaning making. Jeanne Fahnestock's (2003) numerous scientific examples in her analysis of visual and verbal parallelism reinforce the importance of the visual mode for scientific discourse. She finds, for example, that "tabular presentation of instances, examples, or data sets that would otherwise require parallel or repetitive phrasing are the norm in scientific discourse" (Fahnestock, 2003, p. 140).

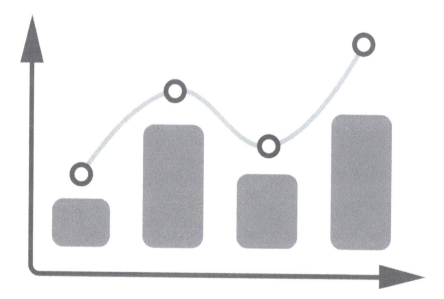

Fig. 35.1. Graphical representations of meaning commonly used in scientific reports. Illustration by Pietrasagh via Wikimedia Commons. https://commons.wikimedia.org/wiki/File:Graph-bar-line-fallback-illustration.svg

In their work on multimodal semiotic analysis, Anthony Baldry and Paul Thibault (2010) examine "meaning compression" in scientific writing, arguing that "scientific texts have always combined and integrated language and visual images in the making of the specialist meanings of scientific discourses" (p. 70). These studies provide a foundation to Gunther Kress's (2010) articulation of mode as a semiotic resource, whereby different modes offer distinctive affordances. For instance, as Kress (2000) illustrates:

> Image is founded on the *logic of display in space*; writing (and speech even more so) is founded on the *logic of succession in time*. Image is spatial and nonsequential; writing and speech are temporal and sequential. That is a profound difference, and its consequences for representation and communication are now beginning to emerge in this semiotic revolution. (p. 339, emphases original)

What Kress (2000) has pointed out is part of an obvious phenomenon that humans have always learned to communicate through multiple sign systems or modes, each of which offers a distinctive way of making meaning (Kress & Bezemer, 2008). To this end, Glynda Hull and Mark Nelson (2005) state, "A multimodal text can create a different system of signification, one that transcends the collective contribution of its constituent parts. More simply put, multimodality can afford, not just a new way to make meaning, but a different kind of meaning" (p. 225). This is a crucial concept to bear for understanding the workings and meanings of multimodal design.

References

Baldry, A., & Thibault, P. (2010). *Multimodal transcription and text analysis* (2nd ed.). Equinox.

Bazerman, C. (1981). What written knowledge does: Three examples of academic discourse. *Philosophy of the Social Sciences, 11*(3), 361–387.

Fahnestock, J. (2003). Verbal and visual parallelism. *Written Communication, 20*(2), 123–152.

Hull, G., & Nelson, M. (2005). Locating the semiotic power of multimodality. *Written Communication, 22*(2), 224–267.

Kress, G., & Bezemer, J. (2008). Writing in multimodal texts: A social semiotic account of design for learning. *Written Communication, 25*(2), 166–195.

Kress, G., & van Leeuwen, T. (2006). *Reading images: The grammar of visual design* (2nd ed.). Routledge.

Recommended Resources

Chandler, D. (2022). *Semiotics: The basics* (4th ed.). Routledge.

Harrison, C. (2003). Visual social semiotics: Understanding how still images make meaning. *Technical Communication, 50*(1), 46–60. https://www.ingentaconnect.com/content/stc/tc/2003/00000050/00000001/art00007

Kress, G. (2000). Multimodality: Challenges to thinking about language. *TESOL Quarterly, 34*(2), 337–340.

Kress, G. (2010). *Multimodality: A social semiotic approach to contemporary communication*. Routledge.

Moriarty, S. E. (2002). The symbiotics of semiotics and visual communication. *Journal of Visual Literacy, 22*(1), 19–28. https://doi.org/10.1080/23796529.2002.11674579

See also: Visual Rhetoric, Materiality, Assemblage

PART 2:
PRACTICES

36. Augment/Augmentation

Jacob Greene

Design is often associated with new media, and there has been an increased attention toward reality-shifting or reality-augmenting technologies in designing everyday user experience. To "augment" means to add to something; to increase it in size, scope, ability, and/or number. In contemporary technological contexts, it is a kind of catch-all term for a wide array of emerging mobile and wearable computing devices that make it possible to "augment" our experience of the world through the affordances of networked digital media, from tourists looking up restaurant reviews in a location-based mobile app to marathon runners tracking their heart rates with a smartwatch. Today, the term has become virtually synonymous with "augmented reality" (AR), or emerging spatial computing technologies capable of overlaying digital media in registration with the user's view of the physical world (Azuma, 1997).

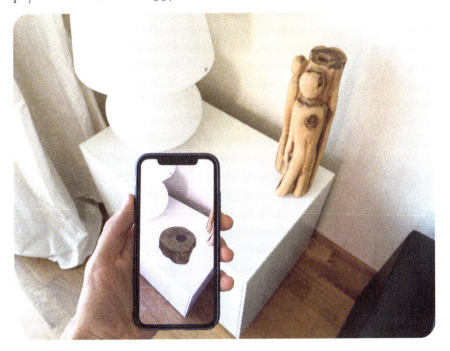

Fig. 36.1. A mobile application that integrates digital objects with physical/real world environments. Photo by UNIBOA on Unsplash. https://bit.ly/keyword-fig-36–1

From a theoretical perspective, the notion of augmentation requires the existence of an entity ontologically stable enough (e.g., a geographic area, a human, "reality" itself, etc.) to be susceptible to qualities of addition. As Casey Boyle and Nathaniel Rivers (2018) point out, this "commonplace conception of augmentation" relies on a "quantitative" approach, which, consequently, reduces the essential rhetorical function of mobile and wearable technologies to their additive characteristics, thus ignoring the "relational and contingent" properties such technologies activate within (and through) specific times and places (pp. 88–89). In other words, the idea of an "augmented reality" must contend with the fact that such technologies do not merely "add on" to a stable notion of reality but actively "transform" it via processes of augmentation (Tinnell, 2011).

For scholars in design and rhetoric and composition, the emergence of technologies of augmentation challenges the relationship between writing, location, and embodiment. Although popular discourses often conceptualize such technologies in terms of bodily improvement (e.g., fitness trackers) and spatial efficiency (e.g., Google Maps), scholars such as Catherine Gouge and John Jones (2016) and Jeff Rice (2012) challenge this discourse of optimization, arguing instead for a more situated, personalized, and creative appropriation of such technologies as rhetorical instruments for complicating preconceived notions of embodiment and spatiality. More recently, Brenta Blevins (2018) has extended many of these insights by delineating the pedagogical potential of augmentation as a process of place-based composing, noting in particular the affordances of mobile technologies for curating historical, cultural, and discursive layers of a location.

Ultimately, critical approaches to this emerging concept must attend to not only the implications of individual technologies but more importantly to how such technologies are reshaping the very political, rhetorical, and ethical frameworks that they purport to "augment."

References

Azuma, R. T. (1997). A survey of augmented reality. *Presence: Teleoperators & virtual environments, 6*(4), 355–385.

Blevins, B. (2018). Teaching digital literacy composing concepts: Focusing on the layers of augmented reality in an era of changing technology. *Computers and Composition, 50,* 21–38.

Boyle, C., & Rivers, N. A. (2018). Augmented publics. In L. Gries and C. G. Brooke (Eds.), *Circulation, writing, and rhetoric*. Utah State University Press.

Gouge, C., & Jones, J. (2016). Wearables, wearing, and the rhetorics that attend to them. *Rhetoric Society Quarterly, 46*(3), 199–206.

Rice, J. (2012). *Digital Detroit: Rhetoric and space in the age of the network*. Southern Illinois University Press.

Tinnell, J. (2011). All the world's a link: The global theater of mobile world browsers. *Enculturation, 12.* http://enculturation.net/all-the-worlds-a-link

Recommended Resources

Crider, J., Greene, J., & Morey, S. (2020). Digital daimons: Algorithmic rhetorics of augmented reality. *Computers and Composition, 57,* 102579. https://doi.org/10.1016/j.compcom.2020.102579

Duin, A. H., & Pedersen, I. (2021). *Writing futures: Collaborative, Algorithmic, Autonomous.* Springer.

Duin, A. H., & Pedersen, I. (2023). *Augmentation technologies and artificial intelligence in technical communication: Designing ethical futures.* Routledge.

Greene, J. (2023). *Composing place: Digital rhetorics for a mobile world.* Utah State University Press.

Kim, Y., & Smith, D. (2017). Pedagogical and technological augmentation of mobile learning for young children interactive learning environments. *Interactive Learning Environments, 25*(1), 4–16. https://doi.org/10.1080/10494820.2015.1087411

Langer, M., & Landers, R. N. (2021). The future of artificial intelligence at work: A review of effects of decision automation and augmentation on workers targeted by algorithms and third-party observers. *Computers in Human Behavior, 123,* 106878. https://doi.org/10.1016/j.chb.2021.10687

See also: Electracy, Digital Rhetoric, Materiality, Environments

37. Bias Toward Action

Katherine Goodman

Although it does not come up regularly in scholarly work, the phrase "bias toward action"—also "bias to action"—is used frequently to describe an attitude in design and making, as well as business (e.g., Amazon, n.d.). This phrase indicates a tendency to understand ideas by building them, test partial prototypes early and throughout the design process, and generally take risks to further specific goals.

Bias toward action is meant to stand in contrast to learning experiences or other work where participants primarily discuss, analyze, and calculate without designing or building anything. Bias toward action can be interpreted as the antonym of the humorous phrase "analysis paralysis," the situation where people become focused on analyzing options to the exclusion of making decisions or actually creating anything.

Design thinking tutorials, including the Stanford d.school Design Thinking Bootleg (2018), often refer to aiming for ideas that are "actionable" or have a step focused on "taking action," even if the exact phrase bias toward action is not used. One example from youth organizations is Girl Scouts USA, which has created "Take Action" badges—earned byscouts when they design and implement original community service projects intended to have a lasting impact (Girl Scouts USA, n.d.). Other phrases in this vein are "make ideas tangible," "thinking through making," and "learning by doing."

Regardless of slogans, the concept promotes the agency of both individuals and teams. The affordances of a design workshops and makerspaces, which include the tools, software, and expertise of other people, are usually intended to support participants in learning, designing, and making. In short, they are environments that support experimentation and tinkering.

References

Amazon. (n.d.). Leadership principles. https://www.amazon.jobs/en/principles

d.school. (2018). Design thinking bootleg. Stanford University. https://dschool.stanford.edu/resources/design-thinking-bootleg

Girl Scouts USA. (n.d.). Community service and take action projects: What's the difference? https://www.girlscouts.org/en/adults/volunteer/

tips-for-troopleaders/programming/community-service-and-take-action-projects.html

Recommended Resources

Bandura, A., & Schunk, D. H. (1981). Cultivating competence, self-efficacy, and intrinsic interest through proximal self-motivation. *Journal of Personality and Social Psychology, 41*, 586–598.

Carlgren, L., Rauth, I., & Elmquist, M. (2016). Framing design thinking: The concept in idea and enactment. *Creativity and Innovation Management, 25*(1), 38–57. https://doi.org/10.1111/caim.12153

Gino, F., Argote, L., Miron-Spektor E., & Todorova, G. (2010). First, get your feet wet: The effects of learning from direct and indirect experience on team creativity. *Organizational Behavior and Human Decision Processes, 111*, 102–115.

Kahneman, D. (2011). *Thinking, fast and slow*. Farrar, Straus and Giroux.

Mishina, K. (1999). Learning by new experiences: Revisiting the flying fortress learning curve. In N. R. Lamoreaux, D.M.G. Raff and P. Temin (Eds.), *Learning by doing in markets, firms, and countries* (pp. 145–184). University of Chicago Press.

See also: Maker Culture, Maker Movement, Invention, Makerspace, Tinkering

38. Brand/Branding

Scott Sundvall

The etymology of "brand" (to brand, branding) can be traced to Old High German (*brinnan*), Old English (*byrnan, biernan, brinnan*), and Middle English (*birnan, brond*), denoting the marking of property with a firebrand, typically a piece of burning wood. The practice of branding, however, precedes the modern term: ceramics were branded in ancient Mesopotamia circa 3,000 BCE, and ancient Egypt branded cattle circa 2,700 BCE.

Corporate marketing provides the modern understanding of branding. Corporate branding is typically a means for lasting name recognition—often by way of uniquely designed mark, emblem, or logo—that also encompasses an ethos or ideology. That is, in corporate terms, the brand identity exceeds the basic commodities the given corporation services, produces, or manufactures.

Fig. 38.1. Vintage Coca-Cola signs at a farmers market. Photo by _ drz _ on Unsplash. https://bit.ly/keyword-fig-38–1

For example, commercials or advertisements selling Coca-Cola or Pepsi do not address the quality of the product commodity. Instead, both corporations rely upon their branding—an idea, a fantasy: Coca-Cola is the conservative soda—the authentic, "real thing"—with commercials pandering to the nostalgic sensibilities of a Norman Rockwell America; Pepsi, on the other hand, casts itself as the "taste of a new generation," with commercials focusing on urbanity, youth, and change. This is the rhetoric and ideology of branding that exceeds the commodities being sold, at least in corporate terms.

The shift from branding as a means to mark property into branding as the marking of property (i.e., commodities) as means to sell such is greatly indebted to visual rhetoric. Commercials and advertisements, which constitute the primary vehicles for advancing a brand identity in our contemporary moment, rely upon a visual and affective dimension of the imaginary, broadly conceived. Insofar as such a rhetorical method is not concerned with the commodity but with the brand identity behind it, such a method draws attention to the "fluff" rather than the "stuff," which Richard Lanham considers to be a basic condition of our emerging attention economy (Lanham, 2007).

The brand, according to the logic of late capitalism, operates by way of basic signals: an idea (ideology) given with the utmost brevity that masks the contents of whatever is actually being given or sold. Additionally, this brevity can be embodied by anything beyond the textual form. Evidently, thanks to the affordance of online interactions, celebrities are paid to be social media influencers, their *selves* having become their own brands. The goal, however, is for rhetoric to learn how to appropriate the logic of branding, as removed from its otherwise exclusive codification by the logic of capital.

Alina Wheeler (2017) notes that "brand identity is tangible and appeals to the senses," (p. 4) or what Gregory Ulmer (2008) refers to as "felt." Jennifer Wingard (2012) argues that this is how brands and brand identities can be rhetorically constructed as *ethos* arrangements when they are actually *pathos* manipulations. Consider how we brand immigrants, for example, with signs of "immigrants crossing"—not too far removed from the original conception of branding. The rhetorical precarity and efficacy of branding can be best demonstrated by Donald Trump: the branding of an otherwise self indicates how the (*sophistic*) fluff can outweigh the (*aletheia*) stuff.

References

Lanham, R (2007). *The economics of attention: style and substance in the age of information*. University of Chicago Press.

Ulmer, G. (2004). *Teletheory*. Atropos Press.

Ulmer, G. (2008). The chora collaborations. *Rhizomes Journal*, Winter, 2008. http://www.rhizomes.net/issue18/ulmer/index.html

131

Ulmer, G. (2019). Electracy: Inventing heuretics. https://heuretics.word-press.com/electracy/

Wheeler, A. (2017). *Designing brand identity: An essential guide for the whole branding team*. Wiley Press.

Wingard, J. (2012). *Branded bodies, rhetoric, and the neoliberal nation-state*. Lexington Books.

Recommended Resources

Bedbury, S. (2003). *A new brand world: eight principles for achieving brand leadership in the twenty-first century*. Penguin Books.

Engbers, S. K. (2013). Branded: the sister arts of rhetoric and design. *Art, Design & Communication in Higher Education, 12*(2), 149–158.

Jacobs, V., & Wintrob, M. (2016). Manifestations of design: Brand, rhetoric, and trends. In N.W. Nixon (Ed.), *Strategic design thinking: Innovation in products, services, experiences, and beyond* (pp. 43–61). Bloomsbury.

Lucas, P. (2014). The rhetoric of brands: How value is generated without substance. *International Journal of Integrated Marketing Communications, 6*(2), 18–24.

Lury, C. (2004). *Brands: The logos of the global economy*. Routledge.

See also: Postdigital Aesthetics, Aesthetics, Visual Rhetoric, Visual Semiotics

39. Coding

Charles Woods

Coding is a foundational practice within the discipline of computer science. As a keystone activity within the maker culture, coding can also be integrated into the classroom in meaningful ways (Vie, 2008). However, some beginners feel nervous when they hear about coding because it may sound highly technical or difficult to learn. However, coding simply means using specific programmatic language to instruct a computer to complete a task. It's like any human language that has its own grammar (rules), structure, and convention (what's typically acceptable and what's not), and it is used to communicate with a computer system.

In movies and pop culture, codes are usually represented in lines of 1s and 0s on a screen (think: *The Matrix*), which is binary coding. Binary coding is a system consisting of a series of 0 and 1 to represent a digit, letter, or character. Binary coding is foundational to the history and practice of coding within computer science, but many other coding languages have since emerged. People who code, often referred to as "coders," use a variety of codes to complete different functions or tasks. For example, the coding language used to produce an online game is different from the codes that are used to create an interactive website. Designers who want to create basic websites will need knowledge of Hypertext Markup Language (HTML) and Cascading Style Sheet (CSS) to structure and display their content. For programming and database building, variations of the C language (C#, C++), Python, Java, and JavaScript may be used.

Here is a bit of a history of codes. In 1972, C, a compiled language program, emerged and has become a comprehensive coding language used in a range of platforms, including Microsoft Windows. An enriched version of C emerged in 1979 when Bjarne Stroustrup developed C++, which adds the object-oriented programming (OOP) paradigm to coding in which programs are organized around objects (or data) rather than logic (i.e., if this/then that) (Kindler and Krivy, 2011).

In the 1980s and 1990s, the demand for personal computing led to the rise of Silicon Valley companies such as Microsoft and Apple. In turn, more coding languages emerged, such as Python in 1991 and JavaScript in 1995. Many coders who started to learn how to code during this time turned to Python—a coding language highlighted by readability with clear syntax—

133

including Google. Of course, one of the distinguishable differences between Web 1.0 and Web 2.0 is interactivity, a keystone rooted in the proliferation of HTML and JavaScript; HTML is used to produce electronic pages, like web pages, and JavaScript is a code for web browsers and interactive websites. In 2020, JavaScript celebrated its twenty-fifth anniversary, and established itself as the most widely used coding language in the world.

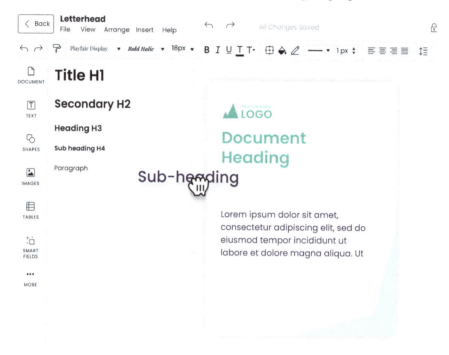

Fig. 39.1. Drag-and-drop functionality afforded by WYSIWYG design tools. Image by Mkotter333 via Wikimedia Commons. https://commons.wikimedia.org/wiki/File:Marq_screenshot_with_drag_and_drop_in_the_editor.png

Coding is a valuable skill for working in the digital age and is also a learnable and applicable skill within maker culture. As WYSIWYG (what-you-see-is-what-you-get) content management platforms like WordPress and Weebly continue to gain popularity as tools that allow users to build interactive websites without learning code, coding still flourishes within the maker community. Instructors who want to teach coding, but are not experts, would be judicious in implementing technologies with which they are familiar (Chapman, 2017) and taking advantage of resources such as Code.org, a practical teaching tool for instructors and students. Girlswhocode.

com is attempting to construct a pipeline of women engineers in the United States through programs that create clear pathways for girls who code in middle and high school. Hourofcode.com allows users to sign up for hour-long coding tutorials in over forty-five languages.

References

Chapman, D. (2017). Designing, implementing, and evaluating a web-based instructional application for technical communication classes. In G. Pullman and B. Gu (Eds.), *Designing web-based applications for 21ˢᵗ century writing classrooms* (pp. 125–140). Baywood Publishing Company, Inc.

Kindler, E., & Krivy, I. (2011). Object-oriented simulation of systems with sophisticated control. *International Journal of General Systems*, *40*, 313–343.

Vie, S. (2008). Digital divide 2.0: "Generation M" and online social networking sites in the composition classroom. *Computers and Composition*, *25*, 9–23.

Recommended Resources

Cox, G., & McLean, A. (2013). *Speaking code: coding as aesthetic and political expression*. MIT Press.

Duin, A. H., & Tham, J. (2018). Cultivating code literacy: Course redesign through advisory board engagement. *Communication Design Quarterly*, *6*(3), 44–58. https://dl.acm.org/doi/abs/10.1145/3309578.3309583

Eyman, D., & Ball, C. E. (2014). Composing for digital publication: Rhetoric, design, code. *Composition Studies, 42*(1), 114–117. https://www.jstor.org/stable/compstud.42.1.0114

Hunt, A., & Thomas, D. (1999). *The pragmatic programmer: From journeyman to master*. Addison-Wesley.

Vee, A. (2017). *Coding literacy: how computer programming is changing writing*. MIT Press.

See also: Digital Rhetoric, Electracy, Data Visualization, Coding Literacy

40. Coding Literacy

Antonio Byrd

Coding literacy refers to understanding computer programming as a political, cultural, material, and social practice that has significant consequences for digital cultures and the lived experiences of coders, clients, and users. This definition draws from foundational research in New Literacy Studies, which argues that reading and writing is not a neutral, valueless skill, but rather a tool infused with and used for ideological ends (Street, 1982; Graff, 1979), whose practices depend on context (Scribner & Cole, 1981). When students learn coding they not only acquire a valuable technical skill or a type of writing (Vee, 2017), they also learn a set of ideological beliefs about coding and software, beliefs that they can embed into the software they create. As manifestations of these values, software or computer instructions can promote the well-being of users, clients, and other coders, or commit violence, especially on communities of marginalized people (Eubanks, 2017; Noble, 2018). Thus, teachers deploying coding in classrooms must be aware of and explicitly discuss the critical frameworks students learn about coding, not just the technical practices and theoretical concepts of making and design thinking.

One ideological framework that persists and gets addressed often when teaching or researching coding literacy is that computer programming centralizes cultures of whiteness and patriarchy (Fancher, 2016; Lin & Besten, 2018). Some community literacy programs that teach coding to achieve social change, i.e., Black Girls Code, The Hidden Genius Project, and others, assist in the upward social mobility of marginalized people, create products that better reflect and situate itself in the lives of marginalized people, and re-write existing narratives that perpetuate stereotypes on women, people of color, and technology. Several computer code bootcamps—career training academies that teach coding in three to six months—recruit and teach people of color programming and workplace skills to graduate on to internships or full time work in software development. Some software developers use their positions and technical skills to create apps that protect or promote the well-being of marginalized communities (Roepe, 2017). In these ways coding is valued for promoting liberation rather than economic and political success for individuals and the nation. These goals are not new but rather continue an ongoing legacy of Black people using technologies like coding for racial justice (Banks, 2005; McIlwain, 2020). The consequences of these efforts, however, vary. For example, a study on African American adults attending a computer code bootcamp for social mobility found that social mobility was hard to come by as students encountered the need to play according to the standards of whiteness in the workplace. Nevertheless, they found alternative literacies or other ways to use coding outside of the workplace for community sustainability and survival (Byrd, 2019).

Although coding offers new ways to teach design- and maker-based curricula, professional and technical communication instructors can encourage students to address values they, and others, attach to coding and programming and to design for social good with an understanding that achieving real change in and beyond the classroom conflicts with institutional and systemic status quo.

References

Banks, A. (2005). *Race, rhetoric, and technology: Searching for higher ground.* Routledge.

Byrd, A. B. (2019). *In pursuit of an uncommon literacy: African American adults' experiences in a computer code bootcamp.* Doctoral dissertation. University of Wisconsin-Madison.

Eubanks, V. (2017). *Automating inequality: How how-tech tools profile, police, and punish the poor.* St. Martin's Press.

Fancher, P. (2016). Composing artificial intelligence: Performing whiteness and masculinity. *Present Tense, 6*(1), 1–8.

Lin, Y. and Besten, M. (2018). Gendered work culture in free/libre open source software development. *Gender, Work, & Organization, 26*(7), 1017–1031. https://doi.org/10.1111/gwao.12255

McIlwain, C. D. (2020). *Black software: The Internet and racial justice, from the Afronet to Black Lives Matter.* Oxford University Press.

Roepe, L. (2017). The software engineer hacking for social change. https://www.ozy.com/rising-stars/the-software-engineer-hacking-for-social-justice/81070.

Scribner, S., & Cole, M. (1981). *The psychology of literacy.* Harvard University Press.

Recommended Resources

Cummings, E. R. (2006). Coding with power: Toward a rhetoric of computer coding and composition. *Computers and Composition, 23*(4), 430–443.

Easter, B. (2018). Feminist_brevity_in_light_of_masculine_long-winded-ness: Code, space, and online misogyny. *Feminist Media Studies, 18*(4), 675–685. https://doi.org/10.1080/14680777.2018.1447335

Ferraio, M., et al. (2014). Software engineering for "social good": Integrating action research, participatory design, and agile development. In P. Jalote, L. Briand, and A. van der Hoek (Eds.), *Companion proceedings of the 36th International Conference on Software Engineering* (pp. 520–523). The Association for Computing Machinery, Inc. https://doi.org/10.1145/2591062.2591121

Fincher, S. A., & Robins, A. V. (Eds.). (2019). *The Cambridge handbook of computing education research (Cambridge Handbook in Psychology)*. Cambridge University Press. doi:10.1017/9781108654555.

Margolis, J., et al. (2008, 2017). *Stuck in the shallow end: Education, race, and computing*. MIT Press.

Noble, S. U. (2017). *Algorithms of the oppressed: How search engines reinforce racism*. New York University Press.

Street, B. V. (1982). *Literacy in theory and practice*. Cambridge University Press.

See also: Coding, Electracy, Hacking, Digital Rhetoric

41. Community Engagement

Sweta Baniya

The maker movement is built on the premise of community engagement. Community engagement in the university context is a collaborative exchange between higher education institutions and the communities they are part of whether it is local, regional/state, national, or international. This kind of collaborative work is often regarded as an exchange between university and community knowledge as well as resources through a stronger partnership that is reciprocal so as to create an impact for both students and the community. Integrating community engagement in academic coursework allows students to directly engage with the community they are part of and understand the community closely with their research as well as interaction. In addition to students learning, the community organization also benefits via this interaction and allows students to use their classroom learning by working on projects that the agency does not have the capacity or skills to do.

In writing studies, community engagement has been a pedagogical approach in which writing students support community-based organizations by implementing the knowledge of writing, research, audience analysis, and document design among others. James Dubinsky (2002) argues that with a course that engages the community, students learn skills they will need in the workplace and gain practical wisdom (*phronesis*) that enables them to be critical citizens. With the experience of his class, Dubinksy (2002) shares that the students in his class have recognized their contribution while understanding community partners' strengths and knowledge and saw this as an opportunity to get involved in their community and work to solve problems. One of the strengths of community-engaged learning is that students get firsthand experiences of how to work in a real-life situation, solve workplace-based issues, and also support the community. This experience-based learning could be rewarding for the students. However, since community engagement projects are additional work that the instructor as well as the students need to do, it sometimes leads to student frustration and lack of engagement.

Additionally, some scholars have critical viewpoints regarding the idea of community engagement pedagogy, such as the educational model being a charity model, treating community as a laboratory, and lack of sustainability in partnership. In this regard, pedagogical approach of commu-

139

nity engagement should be implemented very carefully with the focus on the idea of reciprocity. Robert G. Bringle and Julie A. Hatcher (2002) argue that high quality service-learning classes demonstrate reciprocity between the campus and the community because the service activity is designed and organized to meet both the learning objectives of the course and the service needs identified by the community agency. Reciprocity is very important in community engagement because there should be a win-win situation for both the students, in terms of their learning, and for community partners, in terms of sharing their knowledge and expertise in helping students engage with the community.

References

Bringle, R. G., & Hatcher, J. A. (2002). Campus-community partnerships: The terms of engagement. *Journal of Social Issues, 58*(3), 503–516. https://doi.org/10.1111/1540-4560.00273

Dubinsky, J. M. (2002). Service-learning as a path to virtue: The ideal orator in professional communication. *Michigan Journal of Community Service Learning, 8,* 61–74.

Recommended Resources

Bringle, R. G., & Hatcher, J. A. (2009). Innovative practices in service-learning and curricular engagement. *New Directions for Higher Education, 147,* 37–46. https://doi.org/10.1002/he.356

Bringle, R. G., Hatcher, J. A., & Jones, S. G. (2011). *International service learning: Conceptual frameworks and research.* New York, NY: Routledge.

Cushman, E. (1996). The rhetorician as an agent of social change. *College Composition and Communication, 47*(1), 7–28. https://doi.org/10.2307/358271

Deans, T. (2000). *Writing partnerships: Service-learning in composition.* Urbana, IL: National Council of Teachers of English.

Grabill, J. (2010). Infrastructure outreach and the engaged writing program. In S.K. Rose and I. Weiser (Eds.), *Going public: What writing programs learn from engagement* (pp. 15–28). https://doi.org/10.2307/j.ctt4cgpfh.4

Mayan, M., Lo, S., Richter, S., Dastjerdi, M., & Drummond, J. (2016). Community-based participatory research: Ameliorating conflict when community and research practices meet. *Progress in community health partnerships: Research, education, and action, 10*(2), 259–264. https://doi.org/10.1353/cpr.2016.0023

See also: Cultural Intelligence, Local, Social Innovation, Ethnography

42. Community of Practice

Jeff Naftzinger

Maker communities can be considered a community of practice (CoP). CoP is a group of "people who . . . share a concern or a passion for something they do and learn how to do it better as they interact regularly" (Wenger-Trayner & Wenger-Trayner, 2015, p. 1). These interactions generally involve sharing "experiences, stories, tools, [and/or] ways of addressing recurring problems" (Wenger-Trayner and Wenger-Trayner, 2015, p. 2) about their mutual interest, and in turn learn from each other and improve upon their abilities. This sharing can take place both IRL (in real life) and online, and CoPs can range from more formal and deliberately structured communities, like apprentices learning from experienced tradespeople (Lave & Wenger, 1991), to informal and incidental ones, like coworkers sharing tips and tricks outside of official channels (Brown & Duguid, 2000).

A practice is learned and intentional doing (Wenger, 1998). For example, not all who sew are tailors and not all who cut meat are butchers. It is through learning and engaging with a group of people and developing a shared repertoire—of understandings and techniques—that an activity becomes a practice and a doer becomes a practitioner (Lave and Wenger, 1991). Similarly, members of CoPs tend to have some level of expertise when it comes to their practice, and CoPs require a commitment to learning about and improving upon that expertise. This commitment to learning and improving helps to sustain the practice within the community since participation in the CoP gives members a stake in shaping how it develops (Selfe, 2007).

One of the CoPs most helpful to me as an instructor is my colleagues from graduate school who share an interest in multimodal composing. As graduate students, we regularly interacted with each other in our shared courses, and we also talked with each other in our shared office spaces, in our department's multiliteracy center, and in more formally organized workshops and discussions. Since we were all teaching courses and were all interested in learning more about composition pedagogy, we shared a common practice and we wanted to help each other improve upon that practice. These interactions "provide[d] a space within which to share stories of what works and what does not, learn more about effective strategies for instruction, and expand [our] understanding of technology" (Selfe, 2007, p. 168). As a group, we developed a shared repertoire of assignment sheets, preferred software, suggested lessons, and tips and tricks for teaching. As a new instructor, I was unsure how to include multimodal composing in my classes, but I learned through, and with, my CoP and our shared repertoire. As I became more experienced, I started to help newer members with their multimodal pedagogies. These members, through their participation in the CoP, also helped the rest of us develop and expand our shared repertoire.

Fig. 42.1. Community of practice at work. Photo by KOBU Agency on Unsplash. https://bit.ly/keyword-fig-42-1

Although I am no longer in graduate school, I know that my former department's CoP is still working with each other outside of the academy, and I am still in touch with—and building my repertoire alongside—current and former members online. Through our discussions, our shared repertoire, and our collaborative development, we help each other become practitioners, rather than just doers, of multimodal composing.

References

Brown, J. S., & Duguid, P. (2000). *The social life of information*. Harvard Business Review Press.

Lave, J., & Wenger, E. (1992). *Situated learning: Legitimate peripheral participation.* Cambridge University Press.

Wenger-Trayner, E., & Wenger-Trayner, B. (2015). Communities of practice: A brief introduction. https://wenger-trayner.com/wp-content/uploads/2015/04/07-Brief-introduction-to-communities-of-practice.pdf

Wenger, E. (1998). *Communities of practice: Learning, meaning, and identity.* Cambridge University Press.

Selfe, R. J. (2007). Sustaining multimodal composition. In C. L. Selfe (Ed.), *Multimodal composition: Resources for teachers* (pp. 167–179). Hampton Press.

Recommended Resources

Amin, A., & Roberts, J. (2008). Knowing in action: Beyond communities of practice. *Research Policy, 37*(2), 353–369. https://doi.org/10.1016/j.respol.2007.11.003

Duguid, P. (2005). "The art of knowing": Social and tacit dimensions of knowledge and the limits of the community of practice. *The Information Society, 21*(2), 109–118. https://doi.org/10.1080/01972240590925311

France, H., & Pudelko, B. (2003). Understanding and analysing activity and learning in virtual communities. *Journal of Computer Assisted Learning, 19,* 474–487. https://doi.org/10.1046/j.0266-4909.2003.00051.x

Wenger, E. (2000). Communities of practice and social learning systems. *Organization, 7*(2), 225–246. https://doi.org/10.1177/135050840072002

Zhang, X., & Hu, J. (2022). A study on the learning behaviors and needs of design-maker communities of practice in the era of mobile learning. *Library Hi Tech, 40*(1), 1–27. https://doi.org/10.1108/LHT-12-2021-0486

See also: Radical Collaboration, Shared Leadership, Mentoring

43. Composition Commons

Jess Clements

Composition Commons, like a makerspace, is an alternative nomenclature for contemporary writing centers. While "writing" connotes traditional word-based, print-based methods for scribing letters or characters as visible ideas, words, or symbols on a page, "composition" honors "communicative practice as a dynamic whole" with its "emergent, distributed, historical, and technologically mediated dimensions" as well as its "fundamentally multimodal aspects" (Shipka, 2011, p. 39). "Composition" highlights, in name, a fundamental principle of writing center theory and practice: *process* matters as much or more than *product*. Similarly, "composition" honors the changing nature of twenty-first century writing, which is no longer limited to primarily alphanumeric expression but includes visual, aural, spatial, gestural , and/or linguistic modal choices (Ball, Sheppard, & Arola, 2018).

144

"Center" may inadvertently resonate a picture of writing center practitioners as gatekeepers of privileged university knowledge. "Commons," rather, suggests a lack of privilege or special status, a place or space shared and frequented by the community writ large. Timothy Ballingall (2013) reminds us that early discussions spearheaded by the New London Group that called for shifting writing center ideology away from ivory tower exclusivity and toward more inclusive multiliterate practice were predicated on an interdependence among identity politics, multiliteracy, and multimodality: "the point of multiliteracy . . . is as much about access, difference, and rhetorical agency as it is about text forms" (p. 1). Ballingall suggests writing centers should seek to be culturally literate centers, that writing center practitioners should be invested in bridging discussions of multiliteracy, multimodality, and intersectionality because it will turn the focus away from purchasing expensive hardware to fulfill ambitious multimodal goals and toward leveraging multimodal expression as a means of social justice (p. 2).

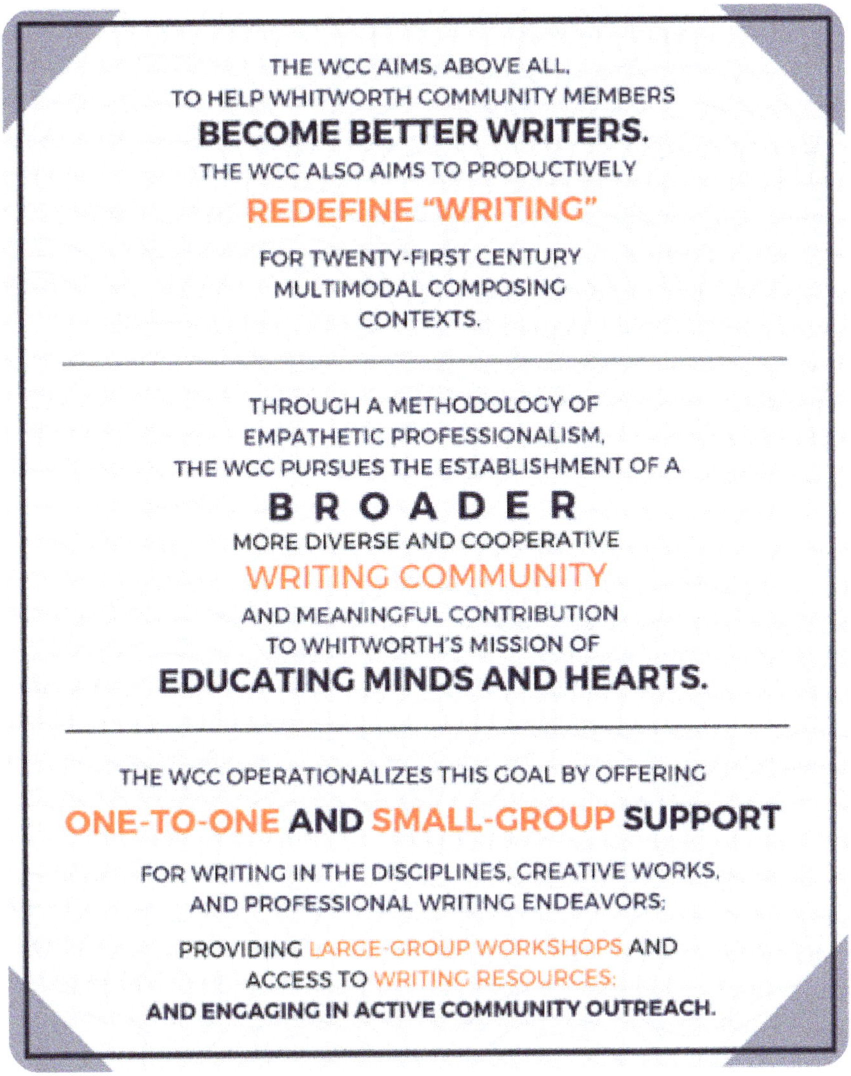

Fig. 43.1. Whitworth Composition Commons (WCC; https://www.whitworth.edu/compositioncommons) mission statement poster. Created by and reproduced with permission from Kalani Padilla.

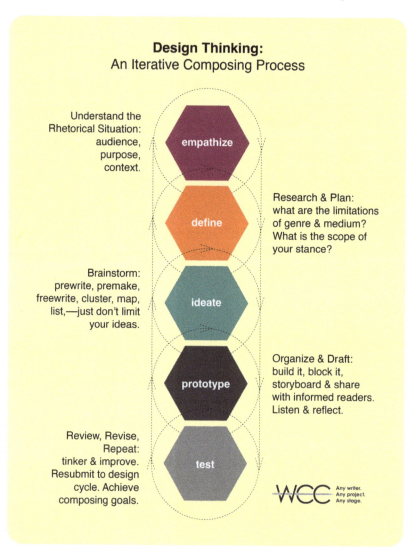

Fig. 43.2. Design Thinking: An Iterative Composition Process. Original infographic by author.

Composition Commons, in name and practice, extends writing centers as multiliterate centers by embracing a practical design-thinking philosophy and makerspace environment for coaching contemporary composing processes as creative problem solving. Iterative engagement is fundamental to design-thinking-based composing as is an embrace of tinkering and a healthy relationship with failure as a natural part of the learning and growing process (Sayers, 2011). The physical expression of a design-centric

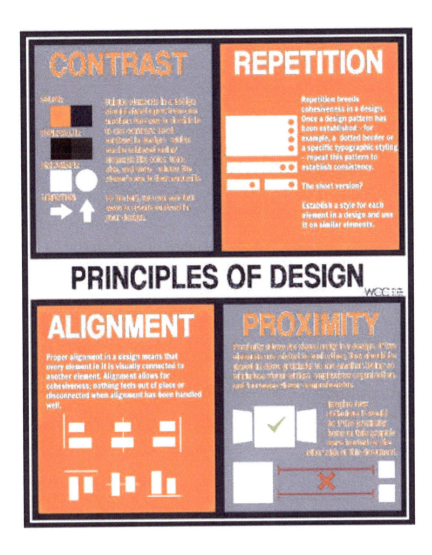

Fig. 43.3. Principles of Design poster. Created by and reproduced with permission from Angela Baik.

thinking environment is a makerspace: a collaborative workspace for making, learning, exploring, and sharing that might use high tech tools—or no tools at all. The physical home of my Composition Commons, for example, includes a two-person computer workstation equipped with access to Adobe Creative Cloud; varying-sized vertically and horizontally anchored white boards with markers and magnets; cork boards and pins with room for prompted and freeform activity expression; Lego tape, blocks, and persons;

literacy games such as Apples to Apples; reference texts; traditional and floor seating with complementary table space; soft and hard lighting options; etc. Wall space is also occupied by inspirational design-thinking-based pedagogical material (an excellent professionalizing activity for Composition Commons consultants!), such as process posters (Fig. 43.2), design principles (Fig. 43.3), and the Commons's mission statement (Fig. 43.1).

Evolution from "writing center" to "composition commons" may include growing pains—convincing key stakeholders, including consultants, co-directors, faculty, and university administrators, that composition commons is more than just a name—but when social justice is at stake and resources abound (see Clements, 2019), the investment is well worth a shift in praxis.

References

Ball, C. E., Sheppard, J., & Arola, K. L. (2018). *Writer/Designer: A guide to making multimodal projects* (2nd ed.). Bedford/St. Martin's.

Ballingall, T. (2013). A hybrid discussion of multiliteracy and identity politics. *Praxis: A Writing Center Journal, 11*(1), 1–6.

Clements, J. (2019). The role of new media expertise in shaping writing center consultations. In K.G. Johnson and T. Roggenbuck (Eds.), *How we teach writing tutors: A WLN digital edited collection*. https://wac.colostate.edu/docs/wln/dec1/Clements.html.

Sayers, J. (2011). Tinker-centric pedagogy in literature and language classrooms. In L. McGrath (Ed.), *Collaborative approaches to the digital in English studies*. Utah State University Press. https://ccdigitalpress.org/book/cad/Ch10_Sayers.pdf

Shipka, J. (2011). *Toward a composition made whole*. University of Pittsburgh Press.

Williams, R. (2015). *The non-designer's design book* (4th ed.). Peachpit Press.

Recommended Resources

Albers, P. (2006). Imagining the possibilities in multimodal curriculum design. *English Education, 38*(2), 75–101. https://www.jstor.org/stable/40173215

Bezemer, J., & Mavers, D. (2011). Multimodal transcription as academic practice: A social semiotic perspective. *International Journal of Social Research Methodology, 14*(3), 191–192. https://doi.org/10.1080/13645579.2011.563616

Dahlström, H. (2021). Students as digital multimodal text designers: A study of resources, affordances, and experience. *British Journal of Educational Technology, 53*(2), 391–407. https://doi.org/10.1111/bjet.13171

Shipka, J. (2009). Negotiating rhetorical, material, methodological, and technological difference: Evaluating multimodal designs. *College Composition and Communication, 61*(1), 343–366.

Walsh, C. S. (2007). Creativity as capital in the literacy classroom: Youth as multimodal designers. *Literacy, 41*(2), 79–85. https://doi.org/10.1111/j.1467–9345.2007.00461.x

See also: Creative Commons Licensing, Makerspace, Open Access

44. Creative Commons Licensing

Quentin Vieregge

Creative Commons licensing is an alternative method of copyrighting intellectual property in comparison to a more traditionally restricted copyright approach. Traditional copyright is intended to encourage the development of new creative works—for instance, books, musical scores, cinematic works, or paintings—by providing developers property rights over their creations (Corbett, 2011, pp. 503–505). However, the expanding importance of collaborative work, peer-production, and sharing online in the 21st century has complicated the efficacy of that system in the eyes of many (Corbett, 2011, pp. 507–509). The Creative Commons project lets artists modulate their intellectual property rights without fully surrendering them (Corbett, 2011, p. 505). For Creative Commons advocates, it is not simply a more practical method but a way to "unlock the full potential of the internet to drive a new era of development, growth and productivity" (Creative Commons, n.d.).

Creators can choose between six different copyrights, each of which allows users slightly different privileges with the creative work in question. Paul Stacey and Sarah Hinchliff Pearson (2017) observe that each of these six licenses has some combination of the following four symbols, each signifying the parameters of the license: BY, SA, ND, and NC (pp. 39–40). *BY* (attribution) means those who use your work must "credit you for the original creation"; *SA* (share-alike) requires them to "license their new creations under identical terms"; *ND* (non-derivative works) means that the work must remain unaltered; and *NC* (non-commercial) means that others can only "build upon your work noncommercially." The most flexible of these licenses only contains the *BY* requirement, requiring merely proper attribution to the original author. The other five licenses have more requirements; for instance, the *BY NC SA* license "lets others remix, tweak, and build upon your work noncommercially, as long as they credit you and license their new creations under the same terms" (Stacey & Pearson, 2017, p. 39–40).

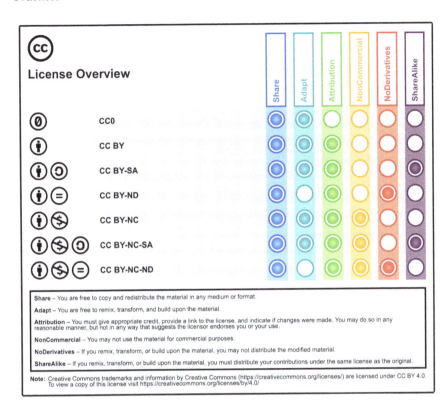

		Share	Adapt	Attribution	NonCommercial	NoDerivatives	ShareAlike
CC0		●	●	○	○	○	○
CC BY		●	●	●	○	○	○
CC BY-SA		●	●	●	○	○	●
CC BY-ND		●	○	●	○	●	○
CC BY-NC		●	●	●	●	○	○
CC BY-NC-SA		●	●	●	●	○	●
CC BY-NC-ND		●	○	●	●	●	○

License Overview

Share – You are free to copy and redistribute the material in any medium or format.

Adapt – You are free to remix, transform, and build upon the material.

Attribution – You must give appropriate credit, provide a link to the license, and indicate if changes were made. You may do so in any reasonable manner, but not in any way that suggests the licensor endorses you or your use.

NonCommercial – You may not use the material for commercial purposes.

NoDerivatives – If you remix, transform, or build upon the material, you may not distribute the modified material.

ShareAlike – If you remix, transform, or build upon the material, you must distribute your contributions under the same license as the original.

Note: Creative Commons trademarks and information by Creative Commons (https://creativecommons.org/licenses/) are licensed under CC BY 4.0. To view a copy of this license visit https://creativecommons.org/licenses/by/4.0/

Fig. 44.1. Creative Commons licensing overview. Graphic by Floba007 via Wikimedia Commons. https://commons.wikimedia.org/wiki/File:CC_License_Overview_Matrix.jpg

When designing a makerspace, Mary Minow, Tomas A. Lipinski, and Gretchen McCord (2016) advise that the curators of that resource familiarize themselves with the basics of copyright law and its practical applications for their institution. For instance, a library would need to determine under what circumstances it—or its employees—would be held liable for copyright violation when its patrons design, create, share, or use objects through the makerspace (Minow, et al., 2016, Q 5.22). One such example would be the use of a 3D printer to design or modify objects. If the maker is using a design that is copyrighted, and that patron's use does not fall under fair use, then they would be engaging in copyright infringement. This could have legal ramifications both for the maker and the library (Minow, et al., 2016, Q. 5.29–30).

There are a number of online resources, such as Flickr, YouTube and MIT OpenCourseWare, that collect intellectual property under a Creative Commons license that makers are encouraged to use (Creative Commons, n.d.). When makers are done creating something in a makerspace, they can even submit their own work with a Creative Commons license; by doing so, they can encourage the spirit of sharing and open source culture that helps inspire the maker movement.

References

Corbett, S. (2011). Creative Commons licenses, the copyright regime and the online community: Is there a fatal disconnect? *The Modern Law Review*, 74, 503–531.

Creative Commons. (n.d.). What we do. https://creativecommons.org/about/

Minow, M., Lipinski, T. A., & McCord, G. (2016). *The library's legal answers for makerspaces* [Kindle version]. https://www.amazon.com/

Stacey, P., & Pearson, S. H. (2017). *Made with Creative Commons*. Ctrl+Alt+Delete Books.

Recommended Resources

Binder, T., Brandt, E., & Gregory, J. (2009). Design participation(-s) – A Creative Commons for ongoing change. *CoDesign, 4*(2), 79–83. https://doi.org/10.1080/15710880802114458

Dobusch, L., & Kapeller, J. (2018). Open strategy-making with crowds and communities: Comparing Wikimedia and Creative Commons. *Long Range Planning, 51*(4), 561–579. https://doi.org/10.1016/j.lrp.2017.08.005

Dobusch, L., & Quack, S. (2010). Epistemic communities and social movements: Transnational dynamics in the case of Creative Commons. In M.-L. Djelic and S. Quack (Eds.), *Transnational communities: Shaping global economic governance* (pp. 226–251). Cambridge University Press.

Garcelon, M. (2009). An information commons? Creative Commons and public access to cultural creation. *New Media & Society, 11*(8), 1307–1326. DOI: 10.1177/1461444809343081

Kim, M. (2007). The Creative Commons and copyright protection in the digital era: Uses of Creative Commons Licenses. *Journal of Computer-Mediated Communication, 13*(1), 187–209. https://doi.org/10.1111/j.1083-6101.2007.00392.x

See also: Open Access, Composition Commons, Intellectual Property

45. Curation

Kathleen Blake Yancey

The word curate, "deriv[ing] from the Latin curare, to take care," includes several interdisciplinary historical associations, among them a ministry's care (as in the "care of souls"), a museum's curatorial processes (McClay, 2018), and a library's collection and cataloging activities. Regardless of its more limited provenance, as Wilfred M. McClay notes, curate is currently employed in multiple contexts—from curating a text to curating the furniture in one's home so it looks like a Hyatt hotel—largely because "curate" elevates practice and object, endowing upon them a "prestige of cultivation, learning, and espirit de finesse" (McClay, 2018). That said, curation is sufficiently stable as concept and practice to inform learning contexts in three ways: in its animation of composing; in its capacity to re-define literacy; and in its (related) adaptability for innovative classroom assignments.

Krista Kennedy (2016), observing that curation is always "rhetorical" (p. 8), points to encyclopedias as one source of curatorial definition. Contrasting the *Cyclopaedia* with *Wikipedia*, Kennedy points to the precedents for curatorial practice created by Chambers's text and to the new forms such practices take: "'new' textual activities" like wikis require both traditional practices—"critical assessment, recomposition, and arrangement of previously disseminated work" —and new authorial practices shifting "the emphasis further from individual originality" (p. 6). The author, in other words, has a different identity in current forms of curation, which Kennedy likens to composing writ large: composing (or designing) itself is curatorial.

Identity, in this case of a self-reflexive learner, is also at the heart of curation in (new) literacy education, although here the focus is twofold: on "authorship with digital media" "across a wide range of sites and media" and on the potential for acting on the world (Potter & Gilje, 2015, p. 2). In this form of literacy education, students curate: "collect[ing], re-arrang[ing] and exhibit[ing] media assets on a wide range of different sites. Consequently, we understand curatorship and learning identity in new media as an emerging literacy practice in which young people's agentive activity is performed *in* and *on* the world" (Potter & Gilje, 2015, p. 2). Interestingly, then, while Kennedy's conception of curation shifts the author away from originality as historically construed, John Potter and Øystein Gilje define curation as author-oriented and agentive.

Three contexts show what such curation might look like in classroom practice. For one assignment, students, creating a visual exhibit using archives and open source tools (e.g., Omeka, Viewshare), engaged in curation defined as "the process by which a series of images or texts is presented to audiences to tell a particular narrative" (Sheffer et al., 2019, p. 80). A second assignment using a similar definition of curation requires students not to draw on existing archival materials, but rather to create them first. Undergraduate interns for the (online) Museum of Everyday Writing locate examples of everyday writing, ranging from letters to memes; upload and create metadata for each; and draw on them to create researched exhibits (Yancey, 2020). And in a third example, students create and curate their electronic portfolios: curation here is defined "as an interpretive act involving three activities: identifying texts; contextualizing texts; and putting texts and contexts into dialogue, or relationship, with each other in a public interface" (Yancey, 2019, p. 141).

References

Kennedy, K. (2016). Textual curation. *Computers and Composition 40*, 175–189.

McClay, W. M. (2018). Curate. *The Hedgehog Review*. https://hedgehogreview.com/issues/the-human-and-the-digital/articles/curate

Potter, J., & Gilje, O. (2015). Curation as a new literacy practice. *Journal of E-learning and Digital Media, 12*(2), 123–127.

Sheffer, J. A., & Hunker, S. D. (2019). Digital curation pedagogy in the archives. *Pedagogy: Critical approaches to teaching literature, language, composition, and culture, 19*(1), 79–107.

Yancey, K. B. (2019). Creating an ePortfolio studio experience: The role of curation, design, and peer review in shaping ePortfolios. In K. B. Yancey (Ed.), *ePortfolio as curriculum: Models and practices for developing students' ePortfolio literacy* (135–149). Sterling, VA: Stylus.

Yancey, K. B. (2020). The Museum of Everyday Writing: Exhibits of everyday writing articulating the past, representing the present, and anticipating the future. *South Atlantic Review, 85*(2). https://museumofeverydaywriting.omeka.net/

Recommended Resources

Finnegan, C. A. (2018). The critic as curator. *Rhetoric Society Quarterly, 48*(4), 405–410. https://doi.org/10.1080/02773945.2018.1479577

O'Neill, P. (2012). *The culture of curating and the curating of culture(s)*. MIT Press.

Synder, I. (2015). Discourses of "curation" in digital times. In R.H. Jones, A. Chik, and C.A. Hafner (Eds.), *Discourse and digital practices: Doing discourse analysis in the digital age* (pp. 209–240). Routledge.

VanHaitsma, P., & Book, C. (2019). Digital curation as collaborative archival method in feminist rhetorics. *Peitho Journal, 21*(2), 505–531.

Yancey, K. B. (2019). *ePortfolio as curriculum: Models and practices for developing students' ePortfolio literacy.* Stylus Publishing.

See also: Assemblage, Memory, Digital Rhetoric, Multigenre

46. Data Visualization

Liz Lane

The decisions one makes when communicating complex information into an engaging and informative visual that can be interpreted by a relatively universal audience are often referred to as data visualization. Data visualization allows one to compose a concise and understandable visual representation of varied information, a process that includes determining the best medium, design principles, and context for a particular set of data or swath of information. Flexible in nature, data visualization often takes the form of "tables, graphs, maps and even text, whether static or dynamic," as each opportunity "provide some means to see what lies within, determine the answer to a question, find relations, and perhaps apprehend things that could not be seen so readily in other forms" (Friendly & Denis, 2001, n.p.).

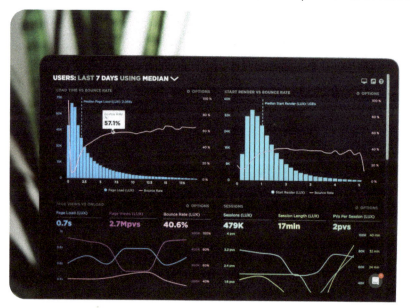

Fig. 46.1. Sample user data visualizations. Photo by Luke Chesser on Unsplash. https://bit.ly/keyword-fig-46–1

Known now as the common shorthand term "data viz," data visualization hearkens back to medieval astronomical drawings such as those from Galileo, statistical graphics, cartography, and ancient languages such as hieroglyphics. (Elkins, 2007; Friendly, 2008). This era's roots are crucial to where we are today in our visual culture, which Friendly describes as an age of "High Data Vis" (Friendly, 2008; Infogram, 2016). As a field, data visualization has grown alongside informatics, human-computer interaction (HCI), and visual rhetoric. As newspapers and the internet helped proliferate the infographic or the graphic explainer (see popular statistics reporting website fivethirtyeight.com for ever-changing examples), data visualization became a more mainstream topic for readers/users in the digital age. We encounter data visualizations in advertisements, on social media, and in most contexts where information, numbers, and figures coalesce.

Noted visual information scholar Edward Tufte defines the term as the practice of communicating information clearly and efficiently (Tufte, 1990). Many in the field of data visualization consider it both an art and a science, as it involves a designer weighing aesthetic choices, medium, labeling, and audience needs in addition to ethical and accurate depictions of information (Elkins, 2007; Friendly, 2008). The potentials of data visualization make the practice highly adaptable, with Tufte noting that "the principles of information design are universal—like mathematics—and are not tied to unique features of a particular language or culture" (Tufte, 1990, p. 10). Accordingly, Tufte offers a schema of "3 Es" for displaying data, which includes confronting ethical, effective, and efficient characteristics of the data with which one is working to best communicate to the reader/viewer.

Seeing and interpreting information are complex activities rife with context and exigence. Data visualization is a dynamic process, one often taken for granted as simply remediating information into a chart or a graph. The designer holds a delicate balance between communicating and conveying, depicting and persuading. Consider the 2018 case of the Thai boy's soccer team trapped in a cave, a harrowing event that offers an excellent case study of data as accessible information in an emergency setting (Sandberg, 2018). During the event, ample data visualizations emerged showing flat and three-dimensional slices of the mountain caves, infographics of geographical location and cavern passageway depths, which included many flooded passageways only skilled divers could navigate.

The challenge in this particular context was for designers to convey the precarity and sheer magnitude of the rescue effort while offering accurate representations of the setting for readers/viewers not immediately in the vicinity of the event (Sandberg, 2018). The story behind the incident was just as crucial to depict as the data governing the rescue of the boys, yet had to be conveyed delicately and accurately to a larger international audience of observers. Data visualization is heavily dependent upon the stories

157

behind information displayed, as the nature of the information and its context is just as important as how it is conveyed to a wide audience or users.

References

Elkins, J. (2007). *How to use your eyes*. Routledge.

Friendly, M. (2008). A brief history of data visualization. In R. M. Haralick (Ed.), *Handbook of data visualization* (pp. 15–56). Heidelberg.

Friendly, M., & Denis, D. J. (2001). Milestones in the history of thematic cartography, statistical graphics, and data visualization. http://www.datavis.ca/milestones/.

Infogram. (2016, June 16). Key figures in the history of data visualization. *Medium*. https://medium.com/@Infogram/key-figures-in-the-history-of-data-visualization-30486681844c

Sandberg, M. (2018, July 05). DataViz as history: Rescue of soccer team trapped for a week in Thailand cave. https://datavizblog.com/2018/07/03/dataviz-as-history-rescue-of-soccer-team-trapped-for-a-week-in-thailand-cave/

Silver, N. (2019). FiveThirtyEight.com. ABC News. https://abcnews.go.com/538

Tufte, E. R. (1990). *Envisioning information*. Graphics press.

Recommended Resources

Gries, L. (2017). Mapping Obama hope: A data visualization project for visual rhetorics. *Kairos: A Journal of Rhetoric, Technology, and Pedagogy, 21*(2). kairos.technorhetoric.net/21.2/topoi/gries/conclusion.html

Hullman, J., & Diakopoulos, N. (2011). Visualization rhetoric: Framing effects in narrative visualization. *IEEE Transactions on Visualization and Computer Graphics, 17*(12), 2231–2240. DOI: 10.1109/TVCG.2011.255

Kostelnick, C. (2008). The visual rhetoric of data displays: The conundrum of clarity. *IEEE Transactions on Professional Communication, 51*(1), 116–130. DOI: 10.1109/TPC.2007.914869

Lupton, E. (2017). *Design is storytelling*. Smithsonian Design Museum. Cooper Hewitt.

Wolfe, J. (2015). Teaching students to focus on the data in data visualization. *Journal of Business and Technical Communication, 29*(3), 344–359. DOI: 10.1177/1050651915573944

See also: Digital Rhetoric, Visual Semiotics, Visual Rhetoric

47. Design Challenge & Makeathon

Jason Tham

The design challenge is a signature activity of the maker movement. It is "an opportunity for designers to flex their critical thinking and problem-solving skills" (JustinMind, 2018). Design challenges come in various scales and expected outcomes. Students in an introduction to product design course may take on a semester-long design challenge, while makers at a community makerspace may participate in a two-week design challenge. In either case, a design challenge aims to tackle complex problems in order to create, or even implement, viable solutions. Activities in a design challenge are informed by the design thinking framework:

- **Discover**: Learn about the people and the context of the problem.
- **Describe**: Synthesize findings from the identified users' points of view.
- **Ideate & prototype**: Conduct iterative ideation and prototyping cycles.
- **Test**: Test ideas and prototypes with actual users.
- **Implement**: Deploy the chosen solution and refine it.

Fig. 47.1. A "GirlsTech" makeathon organized by US Embassy Madrid in 2017. Photograph by US Embassy Madrid via Wikimedia Commons. https://commons.wikimedia.org/wiki/File:Makeathon_ (34286752734).jpg

From an educational standpoint, a design challenge lets its participants focus on the specific tasks that can be managed via effective project management systems and skills, which are essential skills for the technical communication profession. Participants also practice managing available resources, including tools, materials, time, and talent, in order to achieve their project goals. The outcome of this approach is to create an authentic, real-life experience that allows participants to learn by tinkering with tangible, material resources.

A design challenge may take the form of a makeathon—a timed, problem-solving competition designed to spark creativity through radical imagination and collaboration. Many makerspaces host makeathons as a way to engage makers and their communities. While there is not a specific format to follow, a makeathon is typically a one- to two-day event, where participants are given the chance to network and showcase their individual expertise, and form teams through organized or randomized exercises.

For example, the student organizations Tesla Works and Design U at the University of Minnesota have been organizers of the "10,000 Makes" Case Competition, an annual campus-wide makeathon where students of all majors are encouraged to participate. Participants are presented with a case problem the night before the event. According to the makeathon's website,

> 10,000 Makes is an opportunity for students of all colleges to contribute their unique skill sets to tackle real-world issues. This fast-paced one-day event gives students hands-on experience with the design, prototyping, and product pitching process. ("What is 10,000 Makes?" 2019, n.p.)

The goal of a makeathon is to cultivate excitement around designing radical solutions and provide a launchpad to future collaborations. When partnered with community and/or business enterprises, makeathon participants will get the opportunity to extend their marketable ideas to potential funders.

References

JustinMind. (2018). Why you should be doing design challenges. https://www.justinmind.com/blog/why-you-should-be-doing-design-challenges/

"What is 10,000 Makes?" (2019). 10,000 Makes Case Competition. University of Minnesota. https://www.10000makes.com/about

Recommended Resources

"Create Design Challenges Guidelines." (2016). The K12 Lab Wiki. https://dschool-old.stanford.edu/groups/k12/wiki/613e8/Create_Design_Challenges_Guidelines.html

"Frame Your Design Challenge." (n.d.). Design Kit. IDEO. http://www.designkit.org/methods/60

Jones, L. (2018, June 28). A guide to design challenges for product and visual design. *Medium*. https://medium.com/@lukejones/a-guide-to-design-challenges-for-product-visual-design-8ecd53ede143

"Resource: How to organize a makeathon." (n.d.). University Innovation Fellows wiki. https://universityinnovation.org/wiki/Resource:How_to_organize_a_Makeathon

Vermillion, J., & de Salvatierra, A. (2019). Physical computing, prototyping, and participatory pedagogies: Make-A-Thon as interdisciplinary catalyst for bottom-up social change. In *Proceedings of the 2019 eCAADe + SIGraDi Conference: Architecture in the Age of the 4th Industrial Revolution* (pp. 259–366). https://digitalscholarship.unlv.edu/arch_fac_articles/59/

See also: Maker Culture, Maker Movement, DIY, DIWO, Social Innovation

48. DIY/Do It Yourself

Joy Santee

DIY is the "creation, modification, or repair of objects without the aid of paid professionals" (Kuznetsov & Paulos, 2010). DIY encompasses a wide range of practices, from modifying and repurposing IKEA furniture (Rosner & Bean, 2009) to creating cultural artifacts like punk music and zines (amateur, self-published magazines) (Triggs, 2006) to performing at-home brain stimulation (Wexler, 2017). Through DIY, we see a return to pre-commercialized versions of work that value individual and community-based creation of texts and objects for various purposes (Tanenbaum, Williams, Desjardins, & Tanenbaum, 2013).

Motivations for DIY practices vary and may include intrinsic enjoyment of the creativity required of DIY projects, self-sufficiency, personalization of objects or texts for specific purposes, frugality, or political critique of consumerism whereby people create the things they want rather than purchasing them. The benefits of DIY practice to creativity, confidence, and skill development have been extended into education, with pedagogies developed to enhance student learning through DIY making in various fields of study from elementary school through university (Clapp, Ross, Ryan, & Tishman, 2016).

DIY culture is facilitated by the availability of materials, tools, or software used in DIY practice, and DIY practitioners are supported by communities of like-minded people in both online and physical spaces. These communities provide inspiration for new projects, instructional materials for skill development, and social support for individual DIY pursuits. DIY practitioners participate in these communities to learn new ideas and skills, exchange information, or receive social or practical support for their projects (Kuznetsov & Paulos, 2010). Instructables.com is an example of a large online community where participants share instructions for DIY projects and participate in forums to give or receive help. Makeathons and maker faires are examples of events where DIY projects are carried out communally in shared physical spaces. Additionally, print and online publications like *Popular Mechanics* and *Make:* magazine support DIY practitioners.

One contemporary example of DIY is the van dwelling community. The great variety in the size and shape of van interiors means that limited off-the-shelf options exist for furnishing vans as living spaces, necessitating the development of customized solutions. People who live in vans, whether by choice or necessity, modify and customize those vans through building furniture, creating off-grid power systems, and developing efficient storage. This community is brought together by the hashtags #vanlife and #vandwelling on social media and through online spaces like the vandwellers forum on Reddit (https://www.reddit.com/r/vandwellers/). The community is characterized by a DIY ethos of customization, sharing of van dwelling knowledge, and iterative improvements of each person's van through DIY projects. The community has also given rise to shared physical spaces, as in events like the Rubber Tramp Rendezvous, Descend on Bend (an Oregon-based van convention), and builder weekends, where community members meet to help each other with their van builds. The DIY culture of the van dwelling community also provides political critique to modern home-bound culture and its lack of financial and physical freedom.

References

Clapp, E., Ross, J., Ryan, J., & Tishman, S. (2016). *Maker-centered learning: Empowering young people to shape their worlds.* Jossey-Bass.

Kuznetsov, S., & Paulos, E. (2010). Rise of the expert amateur: DIY projects, communities, and cultures. In *Proceedings of the 6th Nordic Conference on Human-Computer Interaction: Extending Boundaries* (pp. 295–304). ACM.

Rosner, D., & Bean, J. (2009). Learning from IKEA hacking: "I'm not one to decoupage a tabletop and call it a day." In *Proceedings of the SIGCHI Conference on Human Factors in Computing Systems* (pp. 419–422). ACM.

Tanenbaum, J., Williams, A., Desjardins, A., & Tanenbaum, K. (2013). Democratizing technology: Pleasure, utility and expressiveness in DIY and maker practice. In *Proceedings of the SIGCHI Conference on Human Factors in Computing Systems* (pp. 2603–2612). ACM.

Triggs, T. (2006). Scissors and glue: Punk fanzines and the creation of a DIY aesthetic. *Journal of Design History, 19*(1): 69–83. https://doi.org/10.1093/jdh/epk00

Wexler, A. (2017). The social context of "Do-It-Yourself" brain stimulation: Neurohackers, biohackers, and lifehackers. *Frontiers in Human Neuroscience, 11.* doi:10.3389/fnhum.2017.00224

Recommended Resources

Anderson, C. (2010). In the next industrial revolution, atoms are the new bits. *Wired, 18*, 2. https://www.wired.com/2010/01/ff_newrevolution/

Chidgey, R. (2014). Developing communities of resistance? Maker pedagogies, do-it-yourself feminism, and DIY citizenship. In M. Ratto and

M. Boler (Eds.), *DIY citizenship: Critical making and social media* (pp. 101–113). MIT Press.

Ratto, M., & Boler, M. (Eds.). (2014). *DIY citizenship: Critical making and social media*. MIT Press.

Sipos, R., & Franzl, K. (2021). Tracing the history of DIY and maker culture in Germany's open workshops. *Digital Culture & Society, 6*(1), 109–120. https://doi.org/10.14361/dcs-2020-0106

Snake-Beings, E. (2017). Maker culture and DiY technologies: Re-functioning as a techno-animist practice. *Continuum, 32*(2), 121–136. https://doi.org/10.1080/10304312.2017.1318825

See also: DIWO, Maker Culture, Teardown, Repair, Tinkering

49. DIWO/Do It With Others

Krys Gollihue

DIY is a concept in pop culture where the individual creates, repairs, or otherwise engages in activities that are outside traditional capitalist economies. The movement began as a response to the mass consumerism of the 1950s in North America, finding traction in hobby communities dedicated to home improvement, electronics, and self-help (Smith, 2014). DIY continued to flourish into the 1970s, 1980s, and the early 1990s, where punk rock countercultures took up the call to "do-it-yourself" as a revolutionary act against the ruling capitalist class. The ethics of DIY continue to be present in contemporary Makerspaces and communities, where technical literacy is closely tied to a maker's ability to be self-sufficient and self-reliant (Dougherty, 2005).

"Do-It-With-Others," or DIWO, is an extension of the DIY ethic into the twenty-first century media landscape, suggesting a more networked and collaborative approach to making that connects people across identities and geographical space. First used in 2006 by the Furtherfield Neighborhood art collective, the term "DIWO" described Internet-based collective art production, enabled through Internet platforms like listservs, email, or portals (Furtherfield, n.d.). In Marc Garrett's (2012) analysis of Furtherfield's use of DIWO, he argues that, unlike the kinds of collaboration used in mainstream design and commercial production, DIWO is an explicitly tactical media form aimed at supplanting establishment art and shifting the means of production back to everyday, diverse artists. Garrett writes that the conditions art must adhere to in order to have representation within the mainstream community are often tied to narrow definitions of innovation and consumerism. DIWO functions under three main elements: the ecological or ethical, the social, and the contemporary networks in which artists function in the twenty-first century. Furtherfield's first DIWO project, "Rosalind," was a call for networked art listed on the Netbehavior listserv and produced an exhibit of thirty images that highlighted "the already thriving imaginations of those who use social networks and digital networks on the Internet as a form of distribution" (Furtherfield, 2007).

More recently, DIWO has been co-opted by the very same industries that it originally aimed to work against and in spite of. Technopedia currently defines DIWO as an approach to project development that allows a more general audience to provide input into the creation of an application, website, product, art project, or product, similar to crowdsourcing. Collaboration tends to be a major organizing principle within corporate makerspaces, fabrication labs, and hackerspaces, where communities work together to bring new innovations to market (Burke, 2011). While the maker movement has received criticism for being an exclusionary culture of rugged individualism and consumer-capitalism (Chachra, 2015), makers are increasingly concerned not only with *how* they can DIWO, but *to what end* they can DIWO (Pinto, 2015). Labs like Seattle Attic have created spaces for patrons to make art around consent, and anarchist spaces like Noisebridge in San Francisco carry on the anti-establishment ethic of the Furtherfield projects (O'Sullivan, 2018). While these contemporary spaces are less concerned with networked online media than the original DIWO movement, their concerns around ownership, representation, and justice reflect the same vision for collaborative design.

References

Burke, J. J. (2014). *Makerspaces: A practical guide for librarians*. Rowman & Littlefield Publishers.

Chachra, D. (2015, January 23). Why I am not a maker. *The Atlantic*. https://www.theatlantic.com/technology/archive/2015/01/why-i-am-not-a-maker/384767

Dougherty, D. (Ed.). (2005). *Make:* magazine (Vol. 1). Maker Media.

Furtherfield. (n.d.). DIWO-do it with others: Resource. http://archive.furtherfield.org/projects/diwo-do-it-others-resource

Garrett, M. (2012). DIWO (do-it-with-others): Artistic co-creation as a decentralized method of peer empowerment in today's multitude. https://seadnetwork.wordpress.com/white-paper-abstracts/final-white-papers/diwo-do-it-with-others-artistic-co-creation-as-a-decentralized-method-of-peer-empowerment-in-todays-multitude-diwo-do-it-with-others-artistic-co-creation-as-a-decentralized-method-of-pe/

O'Sullivan, E. (2018). Excellence in the maker movement. *Journal of Peer Production*, *12*(3), 46–50.

Pinto, L. (2015, May 4). Putting the critical (back) into makerspaces. *The Monitor*. https://www.policyalternatives.ca/publications/monitor/putting-critical-back-makerspaces

Smith, C. D. (2014). Handymen, hippies, and healing: Social transformation through the DIY movement (1940s to 1970s) in North America. *Architectural Histories*, *2*(1), 1–10. ·

Recommended Resources

Caldwell, G. A., & Foth, M. (2014). DIY media architecture: Open and participatory approaches to community engagement. In P. Dalsgaard and A.F. gen Schieck (Eds.), *MAB '14: Proceedings of the 2nd Media Architecture Biennale Conference: World Cities* (pp. 1–10). https://doi.org/10.1145/2682884.2682893

Engchuan, R. N. (2021). Situated assemblages of un-situated things. *Afterall: A Journal of Art, Context and Enquiry, 51*, 36–47. https://www.journals.uchicago.edu/doi/full/10.1086/717399?casa_token=sqif56Xzido-AAAAA:tqryTjbhFmuzqObBhxJZ5UvezJhamUFdd6SaA11b5FCU-WOdFCRkPPpkGTSd-kuxJZevvKy3pFbZtmw#

Maravihas, S., & Martins, J. S. B. (2017). Tacit knowledge in maker spaces and fab labs: From do it yourself (DIY) to do it with others (DIWO). In D. Jaziri-Bouagina and G.L. Jamil (Eds.), *Handbook of research on tacit knowledge management for organizational success* (pp.). IGI Global. DOI: 10.4018/978-1-5225-2394-9.ch011

Rose, M. (2014). Making publics: Documentary as do-it-with-others citizenship. In M. Ratto and M. Boler (Eds.), *DIY citizenship: Critical making and social media* (pp. 201–213). MIT Press.

"What is do it with others (DIWO)?" Definition from Techopedia. (n.d.). Retrieved from http://www.techopedia.com/definition/28410/do-it-with-others-diwo

See also: Feminist Making, Radical Collaboration, DIY, Maker Competencies

167

50. Extreme Situation

R.J. Lambert

Much like the notion of wicked problems, designers often encounter extreme situations as their design challenge. An extreme situation is a context that creates critical hardship for a person or group of people and thereby elicits transformational mental or emotional responses. An extreme situation may have a large impact and affect vast populations, such as a terrorist attack or nuclear accident (Ahn et al., 2017).

Alternatively, an extreme situation may be smaller in scale, affecting one person or very few people, as in the case of a childhood trauma or health crisis (Boyden & Mann, 2005; Marcus, 2012). Rather than its size or reach, the measure of an extreme situation is its psychological toll and personal transformation, as illustrated in the case of hostage survivors (Chaiguerova & Soldatova, 2013) and school shooting survivors (Lambert, 2022). In common vernacular, words like "disaster," "tragedy," or "trauma" are sometimes used to label different kinds of extreme situations, although it is argued that extreme situations create a higher order of hardship and more personalized psychological distress than other kinds of tragedy (Kijak & Funtowicz, 1982, p. 25).

The term "extreme situation" appears to have gained traction with Bruno Bettleheim (1943) in reference to the psychological toll of Nazi concentration camps on internees, which led to split personalities and the breakdown of group affiliations, among other coping responses. Bettleheim characterized an extreme situation by its "shattering impact on the individual," by its "inescapability," by the "expectation" of its continued and indefinite duration, and by the risk that "one's very life would be in jeopardy at every moment" and that "one was powerless to protect oneself" (p. 111). Using the example of prison inmates, sociologist Anthony Giddens (1984) has employed the related term "critical situation" to describe "radical" and "unpredictable" circumstances "which threaten or destroy the certitude of institutionalized routines" (p. 37). Building on Giddens, Paul Marcus (1999) summarized Bettleheim's work as a significant lesson for "how we in our mass society can protect ourselves, resist, and fight back against the assaults on our autonomy, individuality, and humanity" (p. 3).

Marcus (2012) has written extensively on different cases of extreme situations and emphasizes the creative potential that results from psychological shattering within extreme situations. He describes how his personal health crisis during a colon cancer diagnosis and its aftermath set up a "creative, life-affirming, personal transformation" (p. 42). Counterintuitively, then, extreme situations can transform a life-and-death crisis into positive and productive psychological growth.

Outside of psychology, research on extreme situations is most common in fields like organizational management (Moon et al., 2013) and military leadership (Holenweger et al., 2017), where the focus is on preventing or later managing critical disruptions. Since design thinking offers a humane framework for addressing deeply rooted or seemingly intractable problems, extreme situations invite such an approach.

References

Ahn, J., Guarnieri, F., & Furuta, K. (2017). *Resilience: A new paradigm of nuclear safety. From accident mitigation to resilient society facing extreme situations*. SpringerOpen.

Bettelheim, B. (1943). Individual and mass behavior in extreme situations. *The Journal of Abnormal and Social Psychology, 38*(4), 417–452.

Bettelheim, B. (1979). *Surviving and other essays*. Alfred A. Knopf.

Boyden, J., & Mann, G. (2005). Children's risk, resilience, and coping in extreme situations. In *Handbook for working with children and youth: Pathways to resilience across cultures and contexts*, pp. 3–25. Sage Reference.

Chaigeurova, L., & Soldatova, G. (2013). Long-term impact of terrorist attack experience on survivors [*sic*] emotional state and basic beliefs. *Procedia–Social and Behavioral Sciences, 86*, 603–609.

Giddens, A. (1984). *The constitution of society: Outline of the theory of structuration*. Polity.

Holenweger, M., Jager, M. K., & Kernic, F. (2017). *Leadership in extreme situations*. Springer.

Kijak, M., & Funtowicz, S. (1982). The syndrome of the survivor of extreme situations—definitions, difficulties, hypotheses. *International Review of Psycho-Analysis, 9*, 25–33.

Lambert, R. J. (2022). Write or flight in extreme situations: Instability, creativity, and healthy risks. In S. P. Alvarez, Y. Kuchirko, M. McBeth, M. Tarafdar, & M. Watson (Eds.), *Learning and literacy in times of crisis: Emergent teaching through emergencies*. Peter Lang.

Leon, G. R., Kanfer, R., Hoffman, R. G., & Dupre, L. (1991). Interrelationships of personality and coping in a challenging extreme situation. *Journal of Research in Personality, 25*(4), 357–371.

Marcus, P. (1999). *Autonomy in the extreme situation: Bruno Bettelheim, the Nazi concentration camps and the mass society*. Praeger.

Marcus, P. (2012). *In search of the spiritual: Gabriel Marcel, psychoanalysis and the sacred*. Routledge.

Moon, I.-C., Carley, K. M., & Kim, T. G. (2013). *Modeling and simulating command and control for organizations under extreme situations*. Springer.

Recommended Resources

Bonanno, G. A. (2004). Loss, trauma, and human resilience: Have we underestimated the human capacity to thrive after extremely aversive events? *The American Psychologist, 59*(1), 20–28.

Hafen, M. (2016). Of what use (or harm) is a positive health definition? *Journal of Public Health*, 24, 437–441. https://doi.org/10.1007/s10389-016-0741-8

Norris, F. H., Tracy, M., & Galea, S. (2009). Looking for resilience: Understanding the longitudinal trajectories of responses to stress. Social Science & Medicine, 68(12), 2190–2198.

Pickett, K. E., & Wilkinson, R. G. (2010). Inequality: An under-acknowledged source of mental illness and distress. *British Journal of Psychiatry*, *197*, 426–428.

Schwarz, S. (2018). Resilience in psychology: A critical analysis of the concept. *Theory & Psychology, 28*(4), 528–541. https://doi.org/10.1177/0959354318783584

See also: Cognitive Dissonance, Embodiment, Perseverance, Safety

51. FabLab

Estee Beck

The development of FabLabs has inspirational beginnings. MIT Professor Neil Gershenfeld first recognized the potential for tinkering and tactile modalities in the late 1990s when he began teaching, "How to Make (Almost) Anything" at MIT in 1998. It was his goal to introduce engineering students to industrial-size machines for prototyping. In this course, Gershenfeld realized students from multiple disciplines enrolled to learn how to develop and prototype fabricated materials on demand. In seeing this surge of popularity, Gershenfeld collaborated with Bakhtiar Mikhak to create a fabrication laboratory housing industrial tools (Epilog lasers, Trotec lasers, GCC lasers, a ShopBot, 3D printers and scanners, a mini-mill, sewing equipment, large format printing, and more) and the needed computer software to run the hardware.

Fig. 51.1. A Fablab at Umeå University, Sweden. Photograph by Ulrika Bergfors Kriström via Wikimedia Commons. https://commons.wikimedia.org/wiki/File:Sliperiet_Fablab_Ume%C3%A5_University_2.jpg

Along with the hard- and software of the lab, Gershenfeld and Mikhak emphasized additional characteristics to qualify a space as a Fab-Lab—defined by the Fab Foundation as, "a technical prototyping platform for innovation and invention, providing stimulus for local entrepreneurship" (Fab Foundation, 2019, para. 1). These characteristics include public access to the lab, support and subscription to the FabLab charter, and participation in the larger FabLab global network for collaboration. Key to any FabLab space are the people who collaborate and share information. Additionally, the FabLab is a makerspace—or any space or place offering people the opportunity to use materials, including laboratories, hackerspaces, pop-up making spaces, studios, tech shops, museum or community making spaces, knitting and/or crafting circles, woodworking locations, and DIY clubs. When people create, make, tinker, and—most importantly—collaborate with peers, the configuration is a makerspace.

Emphasis on education and networking were (and still are) essential to the sustainability of a global network of FabLabs. People with little to no experience with rapid prototyping can walk into a FabLab and receive personalized support and training on how to use the materials of the space, while collaborating on project designs. By making education a key characteristic of a FabLab, such support furthers what CEO Mark Hatch of Techshop (a membership-based, open access DIY makerspace) calls the maker movement manifesto. Making, sharing, giving, and learning are the four key traits of the call. People must learn safety with the machines, how to control the processes, and how to distribute knowledge to others in order to empower laypersons to continue tinkering, making, and innovating. Former editor in chief of *Wired* and CEO of 3D Robotics, Chris Anderson (2012) furthers this credo and noted the maker movement has roots in the DIY culture precipitated by sharing information and data online across the globe. By capitalizing on knowledge exchange online and the human need to make and share, FabLabs and the maker movement establishes a model for empowerment and entrepreneurship for everyday people from all backgrounds and skill levels.

References

Anderson, C. (2014). *Makers: The new industrial revolution*. Crown Business.

Fab Foundation. (2019). What is a fab lab? https://www.fabfoundation.org/index.php/what-is-a-fab-lab/index.html

Recommended Resources

Alguezaui S., Filieri R. (2010). Investigating the role of social capital in innovation: Sparse versus dense network. *Journal of Knowledge Management, 14*, 891–909.

Hatch, M. (2014). *The maker movement manifesto: Rules for innovation in the new world of crafters, hackers, and tinkerers.* McGraw-Hill Education.

Ramella, F., & Manzo, C. (2018). Into the crisis: Fab labs—a European story. *The Sociological Review Monographs, 66*(2), 341–364. https://doi.org/10.1177/0038026118758535

van Holm, E. J. (2014). What are makerspaces, hackerspaces, and fab labs? SSRN. http://dx.doi.org/10.2139/ssrn.2548211

von Hippel, E. (2017). *Free innovation.* MIT Press.

Walter-Herrmann, J., & Büching, C. (Eds.). (2013). *Fablab: Of machines, makers, and inventors.* Transcript Verlag.

See also: Makerspace, Maker Movement, Digital Fabrication, Physical Computing

52. Feminist Making

Krys Gollihue

While the maker movement has aimed to provide more opportunities for everyday people to learn, access, and create with technology, many scholars, practitioners, and educators have highlighted the ways that making ignores the lived experiences of people of difference, whether they are racial minorities, gender minorities, members of the LGBTQIA (lesbian, gay, bisexual, transgender/transitioning, queer/questioning, intersex, asexual) community, or disabled people. In their analysis of the maker movement's claims to equity, Shirin Vossoughi, Paula K. Hooper and Meg Escudé (2016) demonstrate that along with a same-is-equal mentality across cultural contexts, makerspaces, FabLabs, and design spaces still operate from contained definitions of making, namely products created with current electronic and Internet-enabled tools that are useful and profitable. As such, making practices such as cultural ingenuity, survivance mechanisms, and art—practices often carried out by people of difference—have little value. While liberatory consciousness is a main tenet of the maker movement, research from Drexel University's ExCITe Center (Kim et al., 2019) suggests that makerspaces' attempts to create intentionally inclusive practices are overshadowed by their unintentionally exclusive practices.

As an answer to the problem of unintentional exclusion, many makerspaces, labs, and maker events are providing space for explicitly feminist making practices, i.e. practices that reflect the experiences of people of difference. This may mean more robust programming in fiber materials; workshops that support women entrepreneurs; staffing that reflects the identities of people traditionally excluded from technological spaces; codes of conduct or membership requirements that affirm people of difference; or accessible infrastructure and educational methods. In their review of feminist makerspaces, Em O'Sullivan (2018) found that there was an overwhelming desire among these spaces to affirm the identities of their patrons and to provide programming that is relevant to their lived experiences.

In the United States, several makerspaces have developed programming that centers feminist making and feminist issues related to making. Prototype PGH in Pittsburgh provides workshops focused on gender-inclusive professional development, including a women's business incubator ("Mission," 2019). Seattle Attic has created resources for makers on the issue of consent and holds workshops so that patrons can create projects that explore some aspect of consent (O'Sullivan, 2018). Mothership Hacker-Moms in Berkeley provides on-site childcare for parents wanting to get involved in making. Outside of the United States, feminist makerspaces in the UK, Brazil, and Nairobi have developed programming specific to the economic and environmental constraints that patrons face. The Remakery in south London supports women makers in finding solutions to waste, and rLab in Reading, England hosts repair clinics to help community members learn how to fix their broken things. The Fab Lab Livre Cidade Tiradentes in São Paulo caters to one of the poorest regions in the city, but does so with a "barefoot making" ethic; social inclusion in the makerspace means supporting the immediate needs of the community (Dias & Smith, 2018). These feminist making practices move away from the "making for making's sake" paradigm of the mainstream maker movement and recognize the traditions of making that can affect meaningful change for the local community.

References

Dias, R., & Smith, J. (2018). Making in Brazil: Can we make it work for social inclusion? *Journal of Peer Production*, *12*(1), 43–59.

Kim, Y. E., Edouard, K., Alderfer, K., & Smith, B. K. (2019). *Making culture: A national study of education makerspaces* (pp. 1–18). Philadelphia: Drexel University ExCITe Center.

Mission. (n.d.). Prototype PGH. https://prototypepgh.com/mission

O'Sullivan, E. (2018). Excellence in the maker movement. *Journal of Peer Production*, *12*(3), 46–50.

Vossoughi, S., Hooper, P. K., & Escudé, M. (2016). Making through the lens of culture and power: Towards transformative visions for educational equity. *Harvard Educational Review*, *86*(2), 206–232. https://doi.org/10.17763/0017-8055.86.2.206

Recommended Resources

Alper, M. (2013, March 18). Making space in the makerspace: Building a mixed-ability maker culture. HASTAC. https://www.hastac.org/blogs/merylalper/2013/03/18/making-space-makerspace-building-mixed-ability-maker-culture

Bail, K. (Ed.). (1996). *DIY feminism*. Allen & Unwin Pty Ltd.

Gerstein, J. (2018, July 1). The myth of neutral makerspaces. User Generated Education. https://usergeneratededucation.wordpress.com/2018/07/01/the-myth-of-neutral-makerspaces/

Klimczak, S., Wallace, A., & Gaskins, N. (2016). Technologies of the heart: Beyond #BlackLivesMatter and toward #MakingLiberation. In *Meaningful making: Projects and inspirations for fab labs + makerspaces.* (Vol. 1). Constructing Modern Knowledge Press.

Ring, J. (2021). *Re-tooling the sisterhood: Conceptualizing "meaningful making" through maker culture, makerspace politics, and feminist "little m" making-as-activism.* Doctoral dissertation. Carleton University, Ottawa, Ontario. https://repository.library.carleton.ca/concern/etds/cf95jc326?locale=en

See also: Embodiment, Inclusive Design, Radical Imagination

53. Growth vs. Fixed Mindsets

Katherine Goodman

A growth mindset is the perception that intelligence and talent can be developed or enhanced through deliberate practice or effort. A fixed mindset is the perception that intelligence and talent are relatively set from birth (or during childhood) and that little can be done to expand them. Those with fixed mindsets who run into challenges do not work as hard, because they perceive they have reached the limits of their abilities. This seems to be true for learners, regardless of whether they have been labeled "smart," "slow," or "average." Note that a growth mindset is not the belief that everyone's untrained abilities are identical, but rather, no matter how intelligent or talented someone is, they can continue to grow and improve. Dale Dougherty, founder of *Make:* magazine and co-founder of maker faires, has written that growth mindset "maps very well to the maker mindset, which is a can-do attitude that can be summarized as 'what can you do with what you know?'"(Dougherty, 2013). With its emphasis on framing skills or knowledge not yet mastered as attainable (e.g., "I don't understand this *yet*"), growth mindset aligns with the goals of design thinking, as they both underscore iteration, persistence, and learning from failed attempts.

Growth and fixed mindsets have come into usage largely due to the work of psychologist Carol Dweck (2007). She found that study participants who were praised for working hard on a set of puzzles were more likely to persist as the level of challenge increased, and in some cases utilized "free time" during the study to review materials that would help in future puzzles. In contrast, participants who were praised for being smart mostly chose not to review the extra materials and some even lied about their scores at the end of the study. In addition, the group praised for effort generally did better on final rounds of puzzles than the group praised for being smart. The internalization of these two ways of seeing oneself as a learner are what Dweck and her colleagues have named growth mindset and fixed mindset.

Obviously, a growth mindset is preferred. For a person with a fixed mindset, every opportunity to learn instead becomes a test of identity. Eventually, some people will avoid public learning situations altogether. In this way, fixed mindset is related to learned helplessness, or the idea that people can be trained to give up more easily when faced with adverse conditions where they have no control of the situation or outcomes (Seligman, 1972).

Although there has been a lack of replication for the original findings (Li & Bates, 2017), growth mindset has become a popular topic in K-12 teacher training, with materials encouraging teachers to put more emphasis on praising student effort rather than innate intelligence (e.g., LaRocca, 2017). Some have compared this emphasis on effort to the notion of "grit" as a measurable construct (Duckworth, Peterson, Matthews, & Kelly, 2007).

Many theories of intelligence propose that it is inherently stable, an idea epitomized by IQ tests which purport to measure a construct of intelligence that is steady over a person's lifetime (Boake, 2003). The concept of growth or fixed mindsets sidesteps the issue of measuring intelligence, by instead seeking to understand and influence learning as motivated behavior.

References

Boake, C. (2003). From the Binet–Simon to the Wechsler–Bellevue: Tracing the history of intelligence testing. *Journal of Clinical and Experimental Neuropsychology, 24*(3), 383–405. https://doi.org/10.1076/jcen.24.3.383.981

Dougherty, D. (2013). The maker mindset. In M. Honey & D. E. Kanter (Eds.), *Design, make, play: Growing the next generation of STEM innovators* (pp. 7–11). Routledge.

Duckworth, A. L., Peterson, C., Matthews, M. D., & Kelly, D. R. (2007). Grit: Perseverance and passion for long-term goals. *Journal of Personality and Social Psychology, 92*(6), 1087–1101. https://doi.org/10.1037/0022–3514.92.6.1087

Dweck, C. (2007). *Mindset: The new psychology of success.* Ballantine.

LaRocca, B. (2017). Growth mindset toolkit. https://www.transformingeducation.org/growth-mindset-toolkit/

Li, Y., & Bates, T. C. (2017). Does growth mindset improve children's IQ, educational attainment or response to setbacks? Active-control interventions and data on children's own mindsets. *(In Pre-Press)*, 1–26. https://doi.org/10.31235/osf.io/tsdwy

Seligman, M. E. P. (1972). Learned helplessness. *Annual Review of Medicine, 23*(1), 407–412. https://www.annualreviews.org/doi/pdf/10.1146/annurev.me.23.020172.002203

Recommended Resources

Kelly, T. (2022). A growth mindset. *Appita Magazine*, 2, 18–20.

Laurell, J., Seitamaa, A., Sormunen, K., Seitamaa-Hakkarainen, P., Korhonen, T., & Hakkarainen, K. (2021). A socio-cultural approach to growth-mindset pedagogy. In E. Kuusisto, M. Ubani, P. Nokelainen, and A. Toom (Eds.), *Good teachers for tomorrow's schools: Purpose, values, and talents in education* (pp. 296–312). Brill.

Ludvik, M. B. (2020). How a growth mindset can open one to a decolonization mindset. *About Campus, 25*(5) 25–30.

Somanath, S., Morrison, L., Hughes, J., Sharlin, E., & Sousa, M. C. (2016, February). Engaging "at-risk" students through maker culture activities. In *Proceedings of the TEI'16: Tenth International Conference on Tangible, Embedded, and Embodied Interaction* (pp. 150–158).

Steier, L., & Young, A. W. (2016). Growth mindset and the makerspace educational environment. Sophia, the St. Catherine University repository. https://sophia.stkate.edu/maed/196

See also: Perseverance, STEM & STEAM, Maker Competencies

54. Hacking

Sergio C. Figueiredo

Hacking refers to an arts and humanities practice deeply engaged in technological invention and innovation working across various genres (Levy, 1984/2010). Hackers work to find flaws in systems and to expose those flaws in the name of social, cultural, technological, etc. progress. According to Jan Holmevik (2012), hackers are "present-day *bricoleurs* whose skills and heuretic methodologies [help] subvert and reinvent the very process of digital invention" [emphasis original]. Hackers are makers, designers, and inventors focused on improving the conditions of well-being for individuals and collectives, or at least challenge existing systems in order to create spaces for individuals and collectives to make their own improvements to social well-being.

Etymologically, hacking is defined as a tool and a process of cutting something into pieces (c. 1200), offering connotations with more recent literary-artistic avant-garde practices, such as photomontage strategies (Ulmer, 1983; Losh, 2009). Within the maker movement, hacking includes activities that encourage people to take ownership over the sociotechnical systems that are traditionally controlled by centralized authorities such as technology companies. Often with the goal of repurposing or modifying existing systems and objects to create something new and innovative, hacking emphasizes experimentation, collaboration, and a DIY (do-it-yourself) ethos.

Historically, the literary-artistic avant-garde in the era of the post-Industrial Revolution, as described by Henri Saint-Simon and collaborators in 1825 (Calinescu, 2006), are charged with using the popular arts, like photography, to guide the ethical and social well-being of individuals and collectives living in globally networked societies. Hackers have a similar approach and engagement with digital technologies, grounded in a philosophy whose guiding principles include "sharing, openness, decentralization, and getting your hands on machines at any cost to improve the machines and to improve the world" (Levy, 1984/2010).

However, as a cultural signifier, hacking is a response to the increasing dominance of technology in society. With popular media portraying hackers as criminals who used their skills to break into computer systems for malicious purposes, the negative portrayal has led to increased government regulation and criminalization of hacking. Nonetheless, persistent hackers who believe in the democratization of technology and empowerment of users, new forms of hacking such as hacktivism have been promoted to highlight political and social causes.

      Hackers have been with us all along, even if they have taken on different titles. They are artists, poets, scientists, and activists who take stock of the available means of production and cut up and rearrange those means to create possibilities for new things to emerge out of that work (Losh, 2009, 2012; Wark, 2004), whether it be digital, material, or otherwise. Hacking brings into conversation scientific and technological invention with the traditions of the experimental arts and humanities.

References

Calinescu, M. (2006). *Five faces of modernity: Modernism, avant-garde, decadence, kitsch, postmodernism.* Duke University Press.

Holmevik, J. (2012). *Inter/vention: Free play in the age of electracy.* MIT Press.

Levy, S. (1984/2010). *Hackers: Heroes of the computer revolution.* O'Reilly.

Losh, E. (2012). Hacktivism and the humanities: Programming protest in the era of the digital university. In M.K. Gold (Ed.), *Debates in the digital humanities* (n.p.). Minneapolis, MN: University of Minnesota Press. https://dhdebates.gc.cuny.edu/read/untitled-88c11800-9446-469b-a3be-3fdb36bfbd1e/section/f6fe2a59-8937-4446-a2fb-86dc6ba1975b.

Losh, E. (2009). *Virtualpolitik: An electronic history of government media-making in a time of war, scandal, disaster, miscommunication, and mistakes.* MIT Press.

Ulmer, G. L. (1983). The object of post criticism. In H. Foster (Ed.), *The anti-aesthetic: Essays on postmodern culture* (pp. 83–110). Bay Press.

Wark, M. (2004). *A hacker manifesto.* Harvard University Press.

Recommended Resources

Erickson, J. (2008). *Hacking: The art of exploitation (2ⁿᵈ Edition).* No Scratch Press.

Figueiredo, S. C. (2019). *Hacking the (recommendation) algorithm—an avant-garde provocation.* Textshop Experiments, 6. http://textshopexperiments.org/textshop06/hacking-the-recommendation-algorithm.

Galloway, A., & and Thacker, E. (2007). *The exploit: A theory of networks.* University of Minnesota Press.

Himanen, P. (2002). *The hacker ethic: A radical approach to the philosophy of business.* Random House.

The Mentor. (1986). The conscience of a hacker. *Phrack Magazine*, 1(7). http://phrack.org/issues/7/3.html.

Scheinfeldt, T., & Cohen, D. (Eds.). (2010). Hacking the academy. http://hackingtheacademy.org/.

Ulmer, G. L. (1994). *Heuretics: The logic of invention.* Johns Hopkins University Press.

Vee, A. (2017). *Coding literacy: How computer programming is changing writing.* MIT Press.

Weidman, G. (2014). *Penetration testing: A hands-on introduction to hacking.* No Scratch Press.

See also: Tinkering, Electracy, Repair, Teardown

55. Inclusive Design

Kristen R. Moore

Inclusive design refers to the design of products (technological, communication, or otherwise) that embraces an explicitly inclusive framework or process; conceptually, the design approach is linked with accessibility, disability studies, feminist, anti-racist and/or cultural-critical design. For some, this means universal design, an approach to design that creates the most open and adaptable design of a given product. For the purposes of this entry, however, inclusive design is cast in contrast to traditional approaches to design that are exclusionary for particular groups of people, encouraging designers to consider who has traditionally been at the center of their designs and challenging us to expand our envisioned range of participants and users.

Who is traditionally excluded, and how do we know? Inclusive design considers that systems of oppression have traditionally excluded particular groups from mainstream cultural representation and positions of power in the United States. Inclusive designers consider how that exclusion may also impact the development of communication and technology products. By looking at systematic exclusion (or oppression), inclusive designers consider the ways "the norm" has been constructed as male, white, straight, able-bodied, Christian, and thin. An inclusive design resists the temptation to reproduce the norm by designing for this portrait of a user or participant, instead expanding the design to be more inclusive.

An exclusionary design approach dominates the development of most technologies and communication products, in part because designers themselves tend to reflect what Audre Lorde (2012) termed the "mythical norm" and in part because the mythical norm exists as an unconscious framework for design. For example, recently, the auto industry was criticized for its nearly sole focus on the safety of male drivers, at least in early models: women are forty-seven percent more likely to be injured in vehicle accidents, even when using safety belts. The cause of this safety disparity seems to be that safety tests are run using male (rather than female) metrics and test dummies. Though more recent models have begun practicing inclusive design that considers women as well as men, this example demonstrates the point: the exclusive focus on men created products that are less safe for women.

This example demonstrates the central principle of inclusive design: by acknowledging the ways particular groups of people have been marginalized or excluded historically and systematically, we can attune our design processes to be *more* inclusive. Inclusive design acknowledges that not all designs require consideration of all aspects of the mythical norm; in the above example, the religion of the person is unlikely to affect the successful use or design. However, in the design of a school lunch menu, religion is more likely to affect its exclusionary measures than gender. In this way, inclusive design localizes the who and what of inclusion, asking how the design of a particular technology, process, or product might be predisposed to exclusion based upon its context of use and content.

Inclusion does not rely merely on representation—it's not enough to simply diversify the design team or personas used in development. Instead, inclusive design identifies systems of oppression relevant to the particular technology, process, or product being designed and then intentionally resists those oppressive elements.

Reference

Lorde, A. (2012). *Sister outsider: Essays and speeches.* Crossing Press.

Recommended Resources

Abascal, J., & Nicolle, C. (2005). Moving towards inclusive design guidelines for socially and ethically aware HCI. *Interacting with Computers,* *17*(5), 484–505.

Clarkson, J., Coleman, R., Hosking, I., & Waller, S. (2007). *Inclusive design toolkit.* Cambridge Engineering Design Centre.

Holmes, K. (2018). *Mismatch: How inclusion shapes design.* Cambridge, MA: MIT Press.

Keates, S., Clarkson, P. J., Harrison, L. A., & Robinson, P. (2000, November). Towards a practical inclusive design approach. In *Proceedings on the 2000 Conference on Universal Usability* (pp. 45–52). ACM.

Langdon, P., Clarkson, P. J., & Robinson, P. (Eds.). (2008). *Designing inclusive futures.* Springer Science & Business Media.

Roth, L. (2009). Looking at Shirley, the ultimate norm: Color balance, image technologies, and cognitive equity. *Canadian Journal of Communication,* *34*(1), 111–136.

Shivers-McNair, A. (2017). Localizing communities, goals, communication, and inclusion: A collaborative approach. *Technical Communication,* *64*(2), 97–112.

Shivers-McNair, A., Gonzales, L., & Zhyvotovska, T. (2019). An intersectional technofeminist framework for community-driven technology innovation. *Computers and Composition, 51,* 43–54.

Walton, R., Moore, K., & Jones, N. (2019). *Technical communication after the social justice turn: Building coalitions for action.* Routledge.

See also: Accessibility, Advocacy, Social Innovation, Radical Imagination

56. Information Design

Sarah Welsh

A cell phone bill. Blood test results. A map of weather patterns. A chart of your monthly spending habits. These examples are all the result of careful consideration for the best way to display information so that it is clear to the reader, the user, or even the customer. Information design is the art, science, and practice of presenting information in a way that can be easily and quickly understood.

Fig. 56.1. A user-facing information artifact. Photo by Chanhee Lee on Unsplash. https://bit.ly/keyword-fig-56–1

Information design might encompass graphic design, user experience design, data visualization, infographics, technical writing, and related fields and practices. The term is broadly used to describe communication design and the design of information systems such as wayfinding systems and other dynamic, visual displays (Meirelles, 2013). In short, the main purpose of information design is to inform, which of course covers a broad array of design and related activities.

With the amount of information that is available (online, especially), the importance of professionals who can help a user or reader navigate has become increasingly important. But while information design often goes hand in hand with design of digital spaces, it does have roots in the analog. The use of non-representational images to show numbers only goes back to about the late 1700s—at least in the Western tradition—when Scottish engineer William Playfair wanted to replace numerical tables with something more visually engaging (Tufte, 2001). John Snow was another early pioneer of these methods, and his map of cholera deaths in London is often cited as an early example of the effectiveness of early data visualizations. After plotting and examining a scatter plot map that he constructed, Snow noticed that the illness occurred almost entirely among people who lived near a specific water pump. The pump was shut down, and the epidemic stopped (Tufte, 2001). The practice of visualizing information in this way has of course become increasingly sophisticated with the introduction of digital technology.

Because of the capacity of technology to store such enormous volumes of information, it is even more important that information is organized, efficient, navigable, and able to reach the right audience at the right time (Horn, 2000). Those in the practice of information design consider questions such as: How might a set of data be presented to facilitate communication with a non-expert? How might it be visualized *ethically*? How can you represent complex information clearly to a particular audience, and still do justice to the complexity of that information? How can it be accessible? And while those who are designing information should be looking less to persuade than they are to simply exchange ideas (Jacobson, 2000), information design can always be used to propagandize, to spread false information, or to manipulate. For Jacobson, the person who issues designed information is just as likely as the recipient to be changed by that information. Because information has power, so too does its design.

References

Jacobson, R. (Ed.). (2000). *Information design*. MIT Press.

Horn, R. (2000). Information design: Emergence of a new profession. In R. Jacobson (Ed.), *Information design* (pp. 15–33). MIT Press.

Meirelles, I. (2013). *Design for information*. Rockport Publishers.

Tufte, E. (2001). *The visual display of quantitative information*. Graphics Press, LLC.

Recommended Resources

Cairo, A. (2013). *The functional art: An introduction to information graphics and visualization*. New Riders.

Gibson, D. (2009). *The wayfinding handbook: Information design for public places*. Princeton Architectural Press.

Halpern, O. (2015). *Beautiful data*. Duke University Press.

Montiero, M. (2019). *Ruined by design: How designers destroyed the world, and what we can do to fix it*. Independently published.

Pettersson, R. (2002). *Information design: An introduction*. John Benjamins Publishing Co.

See also: Data Visualization, Interaction Design, User Experience

57. Intellectual Property

Heather Listhartke

Intellectual property encompasses a set of laws that gives rights to artists, makers, writers, and inventors to protect their works from being used by others. The World Intellectual Property Organization (WIPO) calls intellectual property the "creations of the mind, such as inventions; literary and artistic works; designs; and symbols, names and images used in commerce" (n.p.). While many in the US are aware of the laws, many are hazy about what they protect and don't protect. The World Trade Organization divides these rights into two areas of copyright-related rights and industrial property.

Copyrights specifically refer to the "rights of authors of literary and artistic works" which are protected for a minimum of fifty years after the death of the author (World Trade Organization). Industrial property, on the other hand, refers specifically to patents, trademarks, and trade secrets, which each govern specific types of works and have their own specific rules.

Patents, for instance, generally protect things like utility items, new designs for specific utility items, and new varieties of plants (USPTO, 1994). Once the patent is filed for, there's protection of the new thing for twenty years, with potential for adjustment, but these patents only are covered in US-governed areas and can only be enforced by the patentee.

Trademarks are different in that they protect a specific word, name, symbol, or device used in trades of goods or in other words, the branding on a specific item. This specifically protects manufacturers from having a competitor marketing similar items and trying to pass it off as yours, but there are many rules that govern what can and cannot be trademarked like common, everyday phrases or having conflicting similarities. This too, is only enforceable by the one who applied for the trademark (USPTO, 1994).

Finally, under this industrial rights category is trade secrets. This is perhaps the most confusing of the rights as unlike the others, it doesn't have to be something tangible like a product or logo that is being protected, but it could be a specific process of doing something. They also protect things like formulas, patterns, compilations, programs, and devices. The main qualification for this right, according to the US patent and trademark office, is that "it must be used in business, and give an opportunity to obtain an economic advantage over competitors who do not know or use it" (USPTO, 1994)

An example of this would be KFC's classic chicken breading recipe or Coca-Cola's drink formulas. Trade secrets are further different from the above in that they are protected through the World Trade Organization and have different protections beyond simply within the US.

Governing all of these rights is what is commonly referred to as Fair Use, though this is mostly used in copyright instances. As the name suggests, fair use refers to whether the inclusion of another's work is being used fairly. As the University of Texas Libraries points out in their crash course on fair use and copyright, the determination of fair use cases has classically been decided on four factors: 1) What is the character of the use?, 2) What is the nature of the work to be used?, 3) How much of the work will you use?, and 4) What effect would this use have on the market for the original or for permissions if the use were widespread? (Lyon, 2019). More recently though, this has been conflated to two main questions "Is the use you want to make of another's work transformative—that is, does it add value to and repurpose the work for a new audience—and is the amount of material you want to use appropriate to achieve your transformative purpose?" (Lyon, 2019).

Confusion still remains in these cases of fair use, as the determination of "transformative" and "how much" can be subjective according to the specific individual, but many are trying to establish best practices. For instance, the Center for Social Media and Social Impact is working with others to establish codes of best practices for things like working with sound or video or in academic settings.

One piece that has typically not been considered in intellectual property rights or only just being considered in the last decade or so is the collective rights to cultural artifacts or pieces of Native culture since most of the intellectual property rights are written for an individual or specifically named individuals. While there is some work that starts to address this, more needs to be done to address the shortfalls in protecting the pieces that belong to specific groups of people.

References

Lyon, C. (2019, June 18). LibGuides: Copyright crash course: Welcome. http://guides.lib.utexas.edu/copyright

USPTO Office of Public Affairs. (1994, December 01). United States Patent and Trademark Office. https://www.uspto.gov/

World Intellectual Property Organization. (n.d.). About intellectual property. www.wipo.int/about-ip/en/

Recommended Resources

Bennett, P. (2009). Native Americans and intellectual property: The necessity of implementing collective ideals into current United States intellectual property laws. *SSRN Electronic Journal*. doi:10.2139/ssrn.1498783

Brown, M. F. (2003). *Who owns Native culture*. Harvard University Press.

Center for Media and Social Impact. (n.d.). Codes of best practices. https://cmsimpact.org/codes-of-best-practices/

Drahos, P., Mayne, R., & GB, O. (2002). *Global intellectual property rights: Knowledge, access, and development*. Palgrave MacMillan.

Moore, A., & Himma, K. (2018, October 10). Intellectual property. https://plato.stanford.edu/entries/intellectual-property/

Stim, R. (2017, March 25). Fair use. https://fairuse.stanford.edu/overview/fair-use/

Universal Declaration of Human Rights. (n.d.). https://www.un.org/en/universal-declaration-human-rights/index.html

World Trade Organization. (2019). What are intellectual property rights? https://www.wto.org/english/tratop_e/trips_e/intel1_e.htm

See also: Creative Commons Licensing, Digital Rhetoric, Open Access

58. Interaction Design

Jennifer Sano-Franchini

Interaction design (IxD) is an approach to technology design that focuses on how people interact with technologies over time. Such interactions might include how users interact with the technology or the content embedded within it, how the technology facilitates user-user interactions (think: social media), how the technology encourages users to interact with their surroundings in particular ways, or how technologies themselves interact with the environment over time, from production to distribution, to use, to recycling or disposal. Although IxD is applicable to a wide range of technologies, it is often used to talk about the design of digital technologies (Moggridge, 2002; Crampton Smith, 2002; Kolko, 2011). IxD as a term was coined in the mid-1980s by Bill Moggridge and Bill Verplank as a way of reframing how designers think about user experience design (Cooper, Reimann, & Cronin, 2007). That is, IxD is a departure from thinking about technology design solely in terms of functionality or usability, to consider how the technology makes users feel: Is it frustrating? Is it enjoyable? Does it bring pleasure?

To do so, IxD takes an interdisciplinary approach as it draws on design, psychology, art, and emotion to learn about technology users, their physical conditions, their psychology and emotions, and the sociocultural and environmental contexts within which interactions take place. For example, Moggridge (2006) worked with human factors expert Jane Fulton Suri to outline fifty-one methods for "learning about people," including approaches ranging from cognitive task analysis, to historical analysis, to affinity diagramming, to card sorting (pp. 667–681). Others have offered salient considerations for contextualizing these longstanding approaches, however; for instance, Vivianne Castillo (2018) made the convincing case that in order to effectively understand people, designers need to be able to reckon with white cis-male privilege and the biases that come with it, including how the discipline of design is structured to benefit some people over others.

In addition to engaging in research to better understand people, interaction designers often visualize technology use over time through the use of conceptual prototypes (Moggridge, 2006; Pratt & Nunes, 2012) as well as concept maps, process flow diagrams, ecosystem diagrams, and journey maps (Kolko, 2011, pp. 44–50). In centering user interactions, inter-action design theory often attempts to unpack "the micro-level processes by which users physically, cognitively, and emotionally engage and interact with designed objects and experiences" (Sano-Franchini, 2017, p. 87). For example, Dan Saffer (2014) drew attention to these processes through his articulation of "microinteractions," or, "the functional, interactive details of a product" (p. 3), which he suggests are structured in four parts:

- **triggers**, or what the user does to initiate the microinteraction;
- **rules**, or the sequence of actions that take place once the trigger is used;
- **feedback**, or the visual, aural, and/or haptic indicators that helps users understand the rules of the system; and
- **loops and modes**, or the meta-rules that dictate how the system changes, or not, over time.

Designers can draw on these concepts to consider how different parts of a microinteraction work together to reduce complication and obfuscation (Saffer, 2014, pp. 14–22).

Fig. 58.1. Designing a user interaction flow using sketched wireframes. Photo by Kelly Sikkema on Unsplash. https://bit.ly/keyword-fig-58–1

For a full and inclusive understanding of the wide ranging interactions that users engage in as a result of a given technology, it is necessary to consider how different users often engage with the same technology—including its triggers, rules, and feedback—in different ways, whether as a result of physical or psychological differences, or the ways in which they are structurally positioned within our social systems, including on the basis of race, gender, class, sexuality, and/or disability. In this way, IxD should take "into account the potential political, cognitive, and ideological impacts of designed technologies" (Sano-Franchini, 2017, p. 87), including whether particular interactions are equally accessible and available to all users. Interaction designers are often concerned with designing for social good, and must thus be attentive to the potential political, cognitive, and ideological impacts of designed technologies—whether particular interactions uphold or challenge (or both at the same time) dominant or oppressive ideologies, and how social dynamics, including those informed by histories of colonialism, racism, sexism, and ableism inform user interaction and experience. To this end, several organizations and initiatives have been working to advance antiracist and justice-oriented approaches to IxD , such as HmntyCntrd, founded by Vivianne Castillo, the State of Black Design, and the Design Justice Network.

Today, IxD has grown into a significant area of study, with many programs, schools, and organizations across the globe dedicated to the practice and theory of interaction design. Other key concepts for engaging in IxD include goal-oriented design, design as problem solving, IxD as having the potential to help solve "wicked" or large-scale and complex social problems, and usability and user experience research. Through this user-first approach to design, designers are able to come to a stronger understanding of how designs themselves facilitate particular interactions, and the wide ranging implications of those interactions.

References

Castillo, V. (2018). An overdue conversation: The UX research industry's Achilles heel. *HmntyCntrd*. https://www.hmntycntrd.com/resources/an-overdue-conversation-the-ux-research-industrys-achilles-heel

Cooper, A., Reimann, R., & Dubberly, H. (2003). *About face 2.0: The essentials of interaction design*. John Wiley & Sons, Inc.

Crampton Smith, G. (2002). Foreword—What is interaction design? In B. Moggridge (Ed.), *Designing interactions* (n.p.). MIT Press. http://www.designinginteractions.com/chapters/foreword

Kolko, J. (2010). *Thoughts on interaction design*. Morgan Kaufmann.

Moggridge, B. (2006). *Designing interactions*. MIT Press.

Pratt, A., & Nunes, J. (2012). *Interactive design: An introduction to the theory and application of user-centered design*. Rockport Publishers.

Saffer, D. (2013). *Microinteractions: Designing with details*. O'Reilly Media, Inc.

Sano-Franchini, J. (2017). Feminist rhetorics and interaction design: Facilitating socially responsible design. In L. Potts and M.J. Salvo (Eds.), *Rhetoric and experience architecture* (pp. 85–108). Parlor Press.

Recommended Resources

Bardzell, S. (2010, April). Feminist HCI: Taking stock and outlining an agenda for design. In *Proceedings of the SIGCHI Conference on Human Factors in Computing Systems* (pp. 1301–1310). ACM.

Design Justice Network. https://designjustice.org/

Dombrowski, L., Harmon, E., & Fox, S. (2016, June). Social justice-oriented interaction design: Outlining key design strategies and commitments. In *Proceedings of the 2016 ACM Conference on Designing Interactive Systems* (pp. 656–671). ACM.

HmntyCntrd. https://www.hmntycntrd.com/

Interaction Design Association. ixda.org

"Interaction Design Basics." usability.gov

"Interaction Design." Interaction Design Foundation. interaction-design.org/literature/topics/interaction-design

Löwgren, J. (n.d.). Interaction design — brief intro. *The encyclopedia of human-computer interaction, 2nd Ed.* Interaction Design Foundation. https://www.interaction-design.org/literature/book/the-encyclopedia-of-human-computer-interaction-2nd-ed/interaction-design-brief-intro

Löwgren, J., & Stolterman, E. (2004). *Thoughtful interaction design: A design perspective on information technology*. MIT Press.

Moran, K. (2016). The impact of interaction design on brand perception. Nielsen Norman Group. https://www.nngroup.com/articles/interaction-branding/

See also: User Experience, Environments, Embodiment, Wireframing

59. Iterative Design

Cody Reimer

Iterative design (ID) is a process model for design. It is a series of incremental improvements in a development lifecycle accomplished through ideating, modeling, testing, revising, and then repeating that process as necessary. It is linear, cheap, and robust, and one of the oldest foundations for user-centered design (Nielsen, 2011). ID undergirds many disciplines, from human computer interaction (Karat & Karat, 2003) to technical and professional communication (Sullivan, 1989); informs many practices, from methods to pedagogy; and supports many strategies, from participatory design (Spinuzzi, 2005; Taylor, 2006) to agile design (Da Silva et al., 2011). It offers value to practitioners and scholars alike as a means to research, teach, and create.

By iterating with increasingly refined models, designers can move from ideation to validation to delivery with improved confidence in a product's usability. These models take the form of sketches, prototypes, and—in cases such as web and user-interface design—wireframes. While designers have tried to tackle definitional boundaries between sketches and prototypes (Buxton, 2000; Lepore, 2010) to get everyone on the same terms, the common denominator between models is a continuum of fidelity, from low (typically sketches) to high (usually prototypes), with wireframes often in the middle. The continuum of models and fidelity are a gradient with no clear boundaries, only a trajectory toward improvement.

As designers iterate through models, they can test the user experience of each model to aid in subsequent revisions. Lo-fi models are cheap but abstracted, while higher fidelity comes with higher confidence in the user experience data but also a higher resource cost. According to Jakob Nielsen (2011), the recommended number of iterations for a design is at least two (or three versions), but the more the better, he suggests. The limit to the number of iterations is often time, as iterations are cheap compared to alternatives such as parallel design.

The video game industry has benefited from shrinking the time between iterations. As business models shift to longer lifecycles for games, releasing new content for existing software generates significant revenue. That content must be tested within the contexts of the existing game, and the content is on a brisk release schedule. Many video game companies iterate with the speed low-fidelity affords while gaining the confidence high-fidelity permits. This is accomplished by a) tracking user data and, in some cases, releasing that data back to users for them to pore over so as to better enable what in gaming argot is called "theorycrafting" (Paul, 2011), b) employing proxy servers as "test realms" where they can test code without disrupting service, c) stress-testing changes by controlling player access to those test realms, and d) supporting quantitative analysis with qualitative inquiry.

As is the case with many creative processes, ID varies in execution. Different disciplines approach ID differently: ideating (e.g., more democratic, as with participatory design), modeling (sketches, paper prototypes, wireframes, etc.), testing (usability in its myriad forms), and revising are adapted to the contexts and needs of the specific discipline, practice, or strategy employing it.

References

Buxton, B. (2000). *Sketching user experiences: Getting the design right and the right design*. Morgan Kaufman.

Da Silva, T. S., Martin, A., Maurer, F., & Silveira, M. (2011, August). User-centered design and agile methods: A systematic review. In *2011 AGILE Conference* (pp. 77–86). IEEE.

Karat, J., & Karat, C. M. (2003). The evolution of user-centered focus in the human-computer interaction field. *IBM Systems Journal*, 42(4), 532–541.

Lepore, T. (2010, May). Sketches and wireframes and prototypes! Oh my! Creating your own magical wizard experience." *UX Matters*. https://www.uxmatters.com/mt/archives/2010/05/sketches-and-wireframes-and-prototypes-oh-my-creating-your-own-magical-wizard-experience.php

Nielsen, J. (2011, Jan.). Parallel & iterative design + competitive testing = high usability. Nielsen Norman Group. https://www.nngroup.com/articles/parallel-and-iterative-design/

Paul, C. A. (2011). Optimizing play: How theorycrafting changes gameplay & design. *Game Studies, 11*(2). http://gamestudies.org/1102/articles/paul

Spinuzzi, C. (2005). The methodology of participatory design. *Technical Communication, 52*(2), 163–174.

Sullivan, P. (1989). Beyond a narrow conception of usability testing. *IEEE Transactions on Professional Communication, 32*(4), 254–264.

Taylor, T. L. (2006). Beyond management: Considering participatory design and governance in player culture. *First Monday*. http://firstmonday.org/ojs/index.php/fm/article/view/1611/1526

Recommended Resources

Crawford, C. (2002). *The art of interactive design: A euphonious and illuminating guide to building successful software*. No Starch Press.

Goodman, E., Stolterman, E., & Wakkary, R. (2011). Understanding interaction design practices. In *Proceedings of the SIGCHI conference on human factors in computing systems* (pp. 1061–1070). ACM.

Interaction Design Foundation. https://www.interaction-design.org/

Löwgren, J., & Stolterman, E. (2004). *Thoughtful interaction design: A design perspective on information technology*. MIT Press.

Sharp, H. (2003). *Interaction design*. John Wiley & Sons.

Stolterman, E. (2008). The nature of design practice and implications for interaction design research. *International Journal of Design, 2*(1), 55–65. http://ijdesign.org/index.php/IJDesign/article/view/240

See also: Iterative Design, User Requirements, Postmortem

60. Leverage Points

Kristopher Purzycki

In the context of innovation, design thinking prioritizes the identification of leverage points prior to attempting intervention. When analyzing a system with the intent to intervene, leverage points refer to those nodes and processes that, when manipulated even in a marginal fashion, can radically shift the overall construct in a significant way. Even on simple, self-contained structures, the extrapolation of a single, strategic node may produce seismic impacts on those other actants related to it. Nodes made to expansive, multinational ecologies, however, are far more difficult to sufficiently assess. Impacts from changes made to these systems can be difficult to ascertain, potentially resulting in devastating consequences.

To address this, Jay Forrester (1963) argued that organizations must prioritize leverage points according to the measure of impact they have on the overall system. Simply put, intervention in high leverage points yielded quantifiable impacts while low intervention points, which were often those most easily identified and targeted for change, resulted in negligible effect. Environmental scientist Donella Meadows, a colleague of Forrester, expanded the high-low binary into a twelve-category scale. This spectrum of "places to intervene" illustrated a gradation of leverage, from "constants" and "parameters" at the low end to "paradigms" and ideologies at the other. Meadows' (2008) landmark book, *Thinking in Systems: A Primer,* laid out the systems analysis approach she had employed in her previous work *Limits to Growth* and continues to be a foundational work across disciplines.

Due to Meadows' environmental advocacy work, her scale has been lucrative for sustainability studies. Arguing for better identification of "deep" intervention points, David J. Abson et al. (2017) implicated sustainability efforts for failing to identify and engage the true roots of global issues. One well-documented example of this is the model created by Kwamina E. Banson et al. (2014), who demonstrated how better identification of leverage points could help reinforce Ghana's agricultural industry and eliminate export waste. By analyzing several feedback loops (export, agribusiness, quality) that constitute Ghana's international trade, the group were able to identify several conditions that could significantly optimize these complex systems (Banson et al., 2014, p. 683).

Despite its origins in the sciences, the concept of leverage points has also found fertile ground in the humanities. From writing program development (Simpson, 2012) and feminist rhetorical theory (Royster & Kirch, 2012), the concept of leverage points has gained traction in part because of the emergence of the digital humanities in the academy (Rice, 2013).

References

Abson, D. J., Fischer, J., Leventon, J., Newig, J., Schomerus, T., Vilsmaier, U., von Wehrden, H., Abernethy, P., Ives, C. D., Jager, N. W., & Lang, D. J. (2017). Leverage points for sustainability transformation. *Ambio*, *46*(1), 30–39. https://doi.org/10.1007/s13280-016-0800-y

Banson, K. E., Nguyen, N. C., Bosch, O. J., & Nguyen, T. V. (2015). A systems thinking approach to address the complexity of agribusiness for sustainable development in Africa: a case study in Ghana. *Systems Research and Behavioral Science*, *32*(6), 672–688.

Forrester, J. W. (1963). *Industrial dynamics*. MIT Press.

Meadows, D. (2008) *Thinking in systems: A primer.* Chelsea Green Publishing.

Rice, J. (2013). Occupying the digital humanities. *College English*, *75*(4), 360–378

Royster, J. J., & Kirsch, G. E. (2012). *Feminist rhetorical practice: New horizons for rhetoric, composition, and literacy studies.* Southern Illinois University Press. https://ebookcentral.proquest.com

Simpson, S. (2012). The problem of graduate-level writing support: Building a cross-campus graduate writing initiative. *WPA: Writing Program Administration*, *36*(1), 95–118.

Recommended Resources

Forrester, J. W. (1969). *Urban dynamics.* Productivity Press.

Forrester, J. W. (1971). *World dynamics.* Wright-Allen Press.

Gaziulusoy, I., Veselova, E., Hodson, E., Berglund, E., Erdogan Öztekin, E., Houtbeckers, E., Hernberg, H., Jalas, M., Fodor, K., & Ferreira Litowtschenko, M. (2021). Design for sustainability transformations: A deep leverage points research agenda for the (post-)pandemic context. *Strategic Design Research Journal*, *14*(1), 19–31. https://doi.org/10.4013/sdrj.2021.141.02Meadows, D. (1999). *Leverage points: Places to intervene in a system.* The Sustainability Institute.

Penzenstadler, B., Duboc, L., Venters, C. C., Betz, S., Seyff, N., Wnuk, K., Chitchyan, R., Easterbrook, S. M., & Becker, C. (2018). Software engineering for sustainability: Find the leverage points. *IEEE Software*, *35*(4), 22–33. https://doi.org/10.1109/MS.2018.110154908

See also: Social Innovation, Advocacy, Radical Imagination, Inclusive Design

61. Maker Competencies

Estee Beck

The definition of competencies is concerned with skills-based acquisitions of discrete knowledge; however, overlapping descriptions with "literacy" are often used interchangeably with competencies. Development of the initial competencies occurred from January 2016 through August 2016 at the University of Texas at Arlington by the Maker Literacies Task Force, comprised of faculty members and library staff across the disciplines. The specific team members who developed the maker competencies include Martin Wallace, Amanda Alexander, Bonnie Boardman, Morgan Chivers, and Katie Musick Peery. Beta-testers of the initial list of maker competencies included Estee Beck, Bonnie Boardman, Scott Cook, and Cedrick May during the fall 2016 and spring 2017 academic year. The impetus for maker competencies arose from the university providing funding for a FabLab makerspace in the central library on campus, and the need for library staff to identify cross-disciplinary transferable skills for undergraduate and graduate students to acquire and use in coursework and in future employment.

201

There are fifteen core competencies with sub-skills, which are as follows (retrieved from https://library.uta.edu/makerliteracies/competencies):

1. Identify the need to invent, design, fabricate, build, repurpose, repair, or create a new derivative of some "thing" in order to express an idea or emotion, to solve a problem, and/or teach a concept
   a. Recognize unmet needs that may be filled by making
   b. Tinker and hack to learn how things are made and how they work
   c. Evaluate the costs and benefits of making as an alternative to buying or hiring
   d. Investigate how others have approached similar situations

2. Analyze the idea, question, and/or problem
   a. Define the idea, question, and/or problem
   b. Break the idea, question, and/or problem into its constituent parts
   c. Question assumptions

3. Explore the idea, question, and/or problem and potential solutions
   a. Garner input from stakeholders and peers
   b. Research existing relevant products and ideas
   c. Brainstorm a variety of solutions and pursue the most promising

one
  d. Evaluate the costs and benefits of using off-the-shelf parts or kits as opposed to making from scratch

4. Operate safely
  a. Seek training and information on dangerous equipment and materials
  b. Ascertain applicable technical standards and safety codes
  c. Wear personal protective gear when appropriate
  d. Reinforce safety precautions with others
  e. Accustom self with location-specific emergency procedures, egress and disaster plans
  f. Observe safety procedures in the event a person(s) is impaired or injured
  g. Transfer safety principles gleaned in training to broader contexts

5. Assess the availability and appropriateness of tools and materials
  a. Research various equipment and materials to determine limitations and suitability for a specific application
  b. Choose the most appropriate tools and materials (physical, digital, and rhetorical) for the job
  c. Acquire the necessary tools and materials
  d. Investigate alternate tools and materials when a desired tool or material is not available
  e. Fabricate necessary tools, reimagine material choices, develop alternate workflows, and/or revise project scope when alternative tools or materials are not available

6. Produce prototypes
  a. Determine the method of creation most suited to the project
  b. Gain confidence with technologies and processes required for creation
  c. Specify functional requirements for prototype vs desired finished product
  d. Divide design into individual components to facilitate testing
  e. Document design process

7. Utilize iterative design principles
  a. Apply measurable criteria to determine whether creation meets needs
  b. Revise and modify prototype design over multiple iterations
  c. Gather prototype feedback and input from stakeholders and mentors
  d. Rework design to include insights from feedback
  e. Take intelligent risks, use trial and error, and learn from failures

8. Develop a project plan
   a. Identify who the relevant stakeholders are
   b. Specify actionable and measurable project goals and requirements
   c. Utilize time management and project management tools
   d. Outline project milestones, including sequential action items
   e. Anticipate time for multiple prototype iterations
   f. Work effectively within project constraints, be they financial, material, spatial, and/or temporal

9. Assemble effective teams
   a. Recognize opportunities to collaborate with others who provide diverse experiences and perspectives
   b. Gauge the costs & benefits of "doing-it-yourself" (DIY) or "doing-it-together" (DIT)
   c. Recruit team members with diverse skills appropriate for specific project requirements
   d. Join a team where one's skills are sought and valued
   e. Solicit advice, knowledge and specific skills from experts

10. Collaborate effectively with team members and stakeholders
    a. Listen to others
    b. Learn from and with others
    c. Communicate respectfully and clearly with team members and stakeholders
    d. Follow through on team commitments and responsibilities
    e. Practice accountability both personally and with team members
    f. Appraise contributions to the success of the team

11. Employ effective knowledge management practices
    a. Restate technical and maker jargon for the layperson
    b. Document steps clearly with sufficient detail for others to follow and replicate workflows
    c. Use version control to manage project outputs and documentation
    d. Preserve project outputs and documentation for long-term access

12. Apply knowledge gained into other disciplines, workforce, and community
    a. Teach skills and share insights with other makers
    b. Recognize and cultivate transferable skills
    c. Rransfer knowledge, skills, and methods of inquiry across disciplines and activities
    d. Familiarize self with skillsets of others
    e. Connect those seeking to learn something with those who have relevant experience

13. Be mindful of the spectrum of cultural, economic, environmental, and social issues surrounding making
    a. Express awareness of diversity and inclusion when identifying unmet needs
    b. Consider sustainability when making, including upcycling and recycling materials
    c. Scrutinize the ethical implications of making

14. Understand many of the legal issues surrounding making
    a. Demonstrate an understanding of intellectual property rights and protections
    b. Weigh the costs & benefits of seeking intellectual property protections v. making project outputs open and freely available to others
    c. Examine the potential viability of both proprietary and open source systems to adopt/adapt
    d. Respect the intellectual property rights of other makers

15. Pursue entrepreneurial opportunities
    a. Perform thorough market research for competing products and capacity for monetization
    b. Identify project outputs that may be protectable by trade secret, patent, trademark or copyright
    c. Project costs of mass production and requisite economies of scale for return on investment
    d. Refine financial plan for variable scenarios

For more information about maker competencies, including sample lesson plans visit https://library.uta.edu/makerliteracies/curriculum

Recommended Resources

Beck, E. (2020). Discovering maker literacies: Tinkering with a constructionist approach and maker competencies. *Computers and Composition, 58,* 102604. https://doi.org/10.1016/j.compcom.2020.102604

Davidson, A. L., & Price, D. W. (2017). Does your school have the maker fever? An experiential learning approach to developing maker competencies. *LEARNing Landscapes, 11*(1), 103–120.

Hughes, J., & Thompson, S. (2022). Fostering global competencies through maker pedagogies. In J. Hughes (Ed.), *Making, makers, makerspaces: The shift to making in 20 schools* (pp. 57–75). Springer International Publishing.

Wallace, M. K., Trkay, G., Peery, K. M., Chivers, M., & Radniecki, T. (2020). Maker competencies and the undergraduate curriculum. In *ISAM 2018: International Symposium on Academic Makerspaces* (Paper no. 5).

Yoon, J., Kim, K., & Kang, S. J. (2018). Developing maker competency model and exploring maker education plan in the field of elementary and secondary education. *Journal of the Korean Association for Science Education, 38*(5), 649–665.

See also: Maker Movement, Maker Culture, DIWO, Design Challenge & Makeathon

62. Maker Faire

Quentin Vieregge

There are two ways of thinking about maker faire. One is to consider the historical origin and progression of a set of licensed and trademarked gatherings around the world. The second is to consider how these events might inspire teachers, librarians, inventors, hobbyists, and curiosity seekers around the world in their workspaces and community spaces.

Maker faire was the brainchild of Dale Dougherty, founder of *Make:*, a periodical that features people's inventions and gives them guidance on how to build things at home (Dougherty, 2016, pp. 14–15). Through his daily work on his magazine, Dougherty realized there was a latent community of inventors, if only someone could create a forum where they could meet and discuss how they have hacked, modified, or created out of thin air. (Dougherty, 2016, pp. 33–35). Maker faire, which began in San Francisco, California, as an attempt to create such a forum, has reached attendance in the tens and even hundreds of thousands. The festival has spread to major metropolitan areas around the globe such as Washington DC, Tokyo, Lisbon, and even one inside the White House (Loomis, 2016, pp. 20–23).

Fig. 62.1. Scene from a 2018 maker faire in California. Photograph by leighklotz via Wikimedia Commons. https://commons.wikimedia.org/wiki/File:Maker_Faire_(42219774611).jpg

Maker faires are places where inventors showcase their creations and collaborate with other like-minded makers; visitors, including both children and adults, can experiment and learn from interactive exhibits. Dougherty has described the maker faire as "a people's World's Fair that is open to everyone who wants to show their creativity and technical prowess" (2016, p. 34). He sees it as inheriting the long tradition of county faires and international exhibits and emerging out of the participatory culture of the last few decades.

These exhibitions can be sponsored by major corporations advertising their latest creations or by commercial vendors, but the emphasis is on makers—whether professional or amateur—who simply wish to exhibit and share (Maker Media, n.d.). At the 2016 National Maker Faire, which was hosted by the University of the District of Columbia, corporate sponsors, university faculty and students, non-profit groups, and everyday amateurs came together to display a number of exhibits, including movie replicas, robots, and even a "human powered vehicle that was reminiscent of a Formula One race car" (Loomis, 2016, pp. 20–23).

To give some idea of the size and scope of these faires, it is only when they garner at least two hundred exhibits and ten thousand attendants that they are considered a "featured" faire rather than a mini-faire (Maker Media, n.d.). Those who want to create their own maker faires must apply for a license from the company, and if their application is accepted, they will be given guidance on how to market, brand, and operate the maker faire (Maker Media, n.d.).

The financial sustainability of maker faires has been called into question more recently as Maker Media has had to lay off its workforce and shut down its operations, if only temporarily (Constine, 2016). However, it is possible that Maker Media will still license out their brand to others (Constine, 2016). Even if that is not possible, the spirit of maker faire can live on if people simply wish to gather, exchange ideas, and feature their designs.

References

Constine, J. (2016, June). Maker faire halts operations and lays off all staff. *Tech Crunch*. https://techcrunch.com/2019/06/07/make-magazine-maker-media-layoffs/

Dougherty, D., with Conrad, A. (2016). *Free to make: How the maker movement is changing our schools, our jobs, and our minds*. North Atlantic Books.

Loomis, S. (2016, October). Meet your makers: Inside the National Maker Faire. *Techniques, 91*(7), 20–23.

Maker Media. (n.d.). Guidelines. https://makerfaire.com/global/guidelines/

Recommended Resources

Dougherty, D. (2012). The maker movement. *Innovations: Technology, governance, globalization, 7*(3), 11–14. https://direct.mit.edu/itgg/article/7/3/11/9719/The-Maker-Movement

Halverson, E. R., & Sheridan, K. (2014). The maker movement in education. *Harvard educational review, 84*(4), 495–504. https://www.hepg.org/her-home/issues/harvard-educational-review-volume-84-number-4/herarticle/the-maker-movement-in-education

Harlow, D., & Hansen, A. (2018). School maker faires. *Science and Children, 55*(7), 30–37.

Hepp, A. (2018). What makes a maker? Curating a pioneer community through franchising. *Nordisk Tidsskrift for Informationsvidenskab Og Kulturformidling, 7*(2), 3–18. https://tidsskrift.dk/ntik/article/view/111283

Schön, S., Ebner, M., & Kumar, S. (2014). The Maker Movement. Implications of new digital gadgets, fabrication tools and spaces for creative learning and teaching. *eLearning papers, 39*, 14–25. https://core.ac.uk/download/pdf/53025419.pdf

See also: Maker Culture, Maker Movement, DIY, Physical Computing, Tinkering

63. Makerspace

Marijel (Maggie) Melo & Jason Tham

Makerspaces are communal learning environments that center DIY-creation with digital and fabrication tools (Melo & Rabkin, 2019). Makerspaces sometimes are compared to fablabs, hackerspaces, hacklabs or learning labs. The provenance of makerspaces is not exact, but many point to the MIT FabLab as an antecedent to the popularization of the well-known makerspace configuration that has emerged from the maker movement in 2005 ("Fab Lab," 2019). Some say makerspace is a term coined by *Make:* magazine when it was launched in 2005. It became further popularized when the magazine's founder, Dale Dougherty, registered makerspace.com in 2011 and started using the term to refer to open-access spaces for designing and creating (Cavalcanti, 2013).

Fig. 63.1. The SLUB makerspace in Dresden, Germany. Photography by Lukas Boxberger via Wikimedia Commons. https://commons.wikimedia.org/wiki/File:2016_09_UJ_Meet_Up_Makerspace_(c)_Lukas_Boxberger.jpg

Makerspaces often achieve their legibility from a collection of maker-movement specific tools organized within an open learning environment. The Vassar College makerspace defines makerspace as spaces that "combine manufacturing equipment, community, and education for the purposes of enabling community members to design, prototype and create manufactured works that wouldn't be possible to create with the resources available to individuals working alone" ("What is a makerspace," 2015, n.p.). Popular makerspace technologies include fabrication tools (laser cutters, 3D printers, sewing machines), extended reality technologies (augmented, mixed, and virtual reality), robotics, microcomputers and controllers, drones, and materials for prototyping and ideation (Cun, Abramovich, & Smith, 2019).

Makerspaces are typically designed around open and flexible layouts to encourage collaboration among users. Makerspaces are popularly integrated into standalone community centers and/or embedded into library (public, school, and academic) ecosystems. In the US, forty-one percent of state colleges and universities have one or multiple makerspaces (Melo & Rabkin 2019). With increasing growth of makerspaces within universities and other academic institutions, the term "academic makerspaces" is increasingly used to specify a distinguishable field dedicated to studying makerspaces in higher education contexts. To share resources and address makerspace-related problems collaboratively, leading institutions including Yale University, Stanford University, Carnegie Mellon University, and MIT have joined forces to form the Higher Education Makerspaces Initiative (HEMI).

Makerspaces also take on varying forms: digital makerspaces, makerbuses, maker kits (curated kits that are mailed out to users), and pop-up makerspaces. Dana Gierdowski and Dan Reis (2015) show that even a small, mobile makerspace can give students the opportunity to access maker tools at an institution that could not support a campus-wide makerspace.

References

Cavalcanti, G. (2013). Is it a hackerspace, makerspace, techshop, or fablab? *Make:*. https://makezine.com/2013/05/22/the-difference-between-hackerspaces-makerspaces-techshops-and-fablabs/

Cun, A., Abramovich, S., & Smith, J. M. (2019). An assessment matrix for library makerspaces. *Library and Information Science Research*, *41*(1), 39–47.

Melo, M., & Rabkin, A. (2019). Makerspaces in US state universities. https://docs.google.com/spreadsheets/d/1NkcqlAIzTBPuYWt6ynUS1hcFmhhNJQnrvAVtwvZtiCE/edit#gid=1795024749

"Fab Lab." (2019, May 1). Wikipedia. https://en.wikipedia.org/w/index. php?title=Fab_lab&oldid=894991689

Gierdowski, D., & Reis, D. (2015). The MobileMaker: An experiment with a mobile makerspace. *Library Hi Tech, 33*(4), 480–496.

"What is a makerspace?" (2015). VC Makerspace Talks. Vassar College. http://pages.vassar.edu/makerspacetalk/2015/10/21/ what-is-a-makerspace-2/

Recommended Resources

Andrews, M. E., Borrego, M., & Boklage, A. (2021). Self-efficacy and belonging: The impact of a university makerspace. *International Journal of STEM Education, 8*, 1–18. https://link.springer.com/article/10.1186/ s40594-021-00285-0

Eriksson, E., Heath, C., Ljungstrand, P., & Parnes, P. (2018). Makerspace in school—Considerations from a large-scale national testbed. *International Journal of Child-Computer Interaction, 16*, 9–15. https://doi. org/10.1016/j.ijcci.2017.10.001

"Hackerspace." (2019, June 27). Wikipedia. https://en.wikipedia.org/w/ index.php?title=Hackerspace&oldid=903732680

Keune, A., & Peppler, K. (2019). Materials to develop with: The making of a makerspace. *British Journal of Educational Technology, 50*(1), 280–293. https://doi.org/10.1111/bjet.12702

Mersand, S. (2021). The state of makerspace research: A review of the literature. *TechTrends, 65*(2), 174–186. https://link.springer.com/ article/10.1007/s11528-020-00566-5

211

See also: FabLab, Mentoring, Composition Commons, Community of Practice

64. Mentoring

Lyra Hilliard

Mentoring is a fundamental practice in the maker movement and makerspaces. Mentors provide input and guidance to makers who seek advice on their ideas, design, or process. These mentors come in all ages, backgrounds, and experiences. For academics, mentoring is a crucial professional development activity.

Within writing studies, mentoring refers to a broad range of activities that includes orienting rising scholars to their respective disciplines and preparing them to teach other students. Mentoring also extends to professional development for non-tenure track (NTT) and early career faculty. The need for mentoring has been informed by several exigencies, including an increased attention to the quality of undergraduate teaching in general (Stenberg, 2005) and especially the limited number of openings in a competitive job market (Barr Ebest, 2002). This issue has been compounded by a steady decrease in available tenure-track positions and a sharp increase in the use of NTT faculty to teach undergraduate writing courses, both of which exacerbate the stressful conditions under which graduate students and especially NTT faculty labor (Colby & Colby, 2017). In light of this, mentoring is of critical importance not only for individuals' professional development but also for the field's long-term health as a site of vibrant research and exemplary teaching.

Students in doctoral programs are mentored as scholars: they are introduced to the various subfields of and debates within composition and rhetoric (Malenczyk et al., 2018) so that they may join and advance the field with their own research. Graduate students are also mentored as professional academics: they learn how to prepare proposals, present at conferences, develop journal articles, and navigate the job market. Graduate students may also be mentored for work as writing center tutors, writing center administrators, or writing program administrators (Gebhardt & Gebhardt, 2013). Whereas graduate students are mentored in preparation to leave their respective departments, NTT faculty are mentored to stay in their respective departments. This latter type of mentoring, when it is available at all, is almost exclusively focused on teaching (Artze-Vega et al., 2013).

The vast majority of college writing instructors learn how to teach through their graduate programs. The amount of pedagogical mentoring provided to brand new teachers ranges widely across institutions: some offer week-long "boot camps" immediately before the start of the semester; some offer credit-bearing teaching practicum; some offer a mixture of both (Stenberg, 2005). Ongoing pedagogical mentoring for newer teachers and NTT faculty may include pre-semester orientations, bi-monthly mentor group meetings, and monthly workshops on topics such as cultivating an inclusive classroom, teaching with technology, and responding to student writing. Whereas many graduate programs offer a single teaching seminar in composition pedagogy, some offer additional coursework on specializations such as teaching professional writing (Ohio State) and teaching multilingual writers (Arizona State). In addition, some departments offer semester-long mentoring programs for specialized skills such as teaching writing online that may be taken by both graduate students and NTT faculty (U Maryland).

Fig. 64.1. Mentoring can happen on the job when teaching opportunities become available. Photo by Jason Goodman on Unsplash. https://bit.ly/keyword-fig-64–1

Extra-institutional mentoring opportunities are available to new and emerging scholars who attend conferences and engage with academic communities online. Many professional associations have robust mentoring opportunities for graduate students, such as Computers & Writing's Graduate Research Network and the Writing Program Administrator's Graduate Organization. Other associations mentor both graduate students and early-career researchers, such as the Coalition of Feminist Scholars in the History of Rhetoric and Composition (CFSHRC) and the CCCC through its Research Network Forum. Professional scholars of all ranks find additional community through professional listservs and social networks such as X (formerly Twitter).

References

Artze-Vega, I., Bowdon, M., Emmons, K., Eodice, M., Hess, S. K., Lamonica, C. C., & Nelms, G. (2013). Privileging pedagogy: Composition, rhetoric, and faculty development. *College Composition and Communication, 65*(1), 162–184.

Barr Ebest, S. (2002). Mentoring: Past, present and future. In B. P. Pytlik & S. Liggett (Eds.), *Preparing college teachers of writing: History, theories, programs, practices* (pp. 211–221). Oxford.

Colby, R., & Colby, R. S. (2017). Real faculty but not: The full-time, non-tenure-track position as contingent labor. In S. Kahn, W. Lalicker, and A. Lynch-Biniek (Eds.), *Contingency, exploitation, and solidarity: Labor and action in English composition,* (p. 57). The WAC Clearinghouse.

Gebhardt, R. C., & Gebhardt, B. G. S. (2013). *Academic advancement in composition studies: Scholarship, publication, promotion, tenure.* Erlbaum.

Malenczyk, R., Miller-Cochran, S., Wardle, E., & Yancey, K. (Eds.) (2018). *Composition, rhetoric, and disciplinarity.* University Press of Colorado.

Stenberg, S. J. (2005). *Professing and pedagogy: Learning the teaching of English.* NCTE.

Recommended Resources

Anderson, V., & Romano, S. (Eds.). (2006). *Culture shock and the practice of profession: training the next wave in rhetoric and composition.* Hampton Press.

Clark, E., Minton, S., & Andree, J. (2006). Training and mentoring for adjuncts. *Forum: Newsletter of the Non-Tenure-Track Faculty Special Interest Group 10.1 [insert in College Composition and Communication 58.1]*, A12–A14.

Das Bender, G. (2002). Orientation and mentoring: collaborative practices in teacher preparation. In B. P. Pytlik and S. Liggett (Eds.), *Preparing college teachers of writing: history, theories, programs, practices* (pp. 233–241). Oxford.

Dobrin, S. I. (Ed.). (2005). *Don't call it that: The composition practicum.* National Council of Teachers of English.

Martin, W., & Paine, C. (2002). Mentors, models, and agents of change: Veteran TAs preparing teachers of writing. In B. P. Pytlik and S. Liggett (Eds.), *Preparing college teachers of writing: history, theories, programs, practices* (pp. 222–232). Oxford.

Penrose, A. M. (2012). Professional identity in a contingent-labor profession: Expertise, autonomy, community in composition teaching. *WPA: Writing Program Administration–Journal of the Council of Writing Program Administrators, 35*(2), 108–126.

VanHaitsma, P., & Ceraso, S. (2017). "Making it" in the academy through horizontal mentoring. *Peitho: Journal of the Coalition of Feminist Scholars in the History of Rhetoric and Composition, 19*(2), 210–233.

See also: Shared Leadership, Community of Practice, Writing Studio

65. Open Access

Dana Lynn Driscoll

The ownership of ideas, texts, images, music, and many other media are determined by copyright laws. Copyrights rose to prominence in the West as the printing press became a prominent force in the fifteenth and sixteenth centuries and the "right to copy" became a contested issue (Geller, 2000). These movements led to stricter copyright laws, where by the twenty-first century, copyright laws determine much of the intellectual property produced—an individual who is a creator, maker, or publisher often retains the sole rights to produce (and thus profit) from those wanting to access works. This restrictive and profit-oriented philosophy drives much of the current publishing, software, consumer goods, music, and entertainment industry.

The open access movement arose as a response to these severe restrictions and limitations on access and use (Kennedy, 2001). While copyright focuses on restricting access, open access allows works to be freely available for anyone to access, and in some cases, use, modify, and redistribute, depending on the specific license. Called by a variety of different names, each with slightly different connotations, terms associated with this movement include open source, open access, copyleft, open educational resources/open learning, creative commons, and share alike. One of the first examples of this kind of licensing was for computer software, with the 1989 GNU Software License (Stallman, 1989). The most well-known open access license is Creative Commons, and the "share alike" principle, now used globally around the world for software, music, videos, materials, images, blueprints, designs and much more (Goss, 2007).

Within the maker movement, open access is critical to the success of the overall movement. In *The Maker Manifesto*, Mark Hatch (2013) describes the core aspects of the maker movement as including to make, give, share, learn, tool up, play, participate, support, and change. Many of these principles are strongly enhanced through open access. Precious Plastic (https://preciousplastic.com/) offers a great case study of how the maker movement and open access are intertwined. Precious Plastic offers open-source blueprints and video tutorials that teach people how to build and use the tools to recycle plastic at home and turn that plastic into a variety of new goods. A global map of individuals connects people who are creating makerspaces, building tools, or interested in participating. Community forums support conversation and collaboration, and a marketplace allows people who are using the tools to showcase and sell their products. In this case, open sourcing of the blueprints and tutorials, combined with the free expertise of the volunteers, literally drives the movement forward and empowers thousands around the globe to create from waste.

While open access principles are focused on freely sharing, making, and giving, open access does not equal universal access. Nearly all of the open access movement is Internet-driven, which presents barriers. As Peter Suber (2013) notes, technological, linguistic, disability, or censorship barriers may still prevent an individual from accessing the work. For example, someone without access to the Internet (for a variety of reasons) could not benefit or participate in these open access movements.

References

Geller, P. E. (2000). Copyright history and the future: What's culture got to do with it. *Journal of the Copyright Society of the USA, 47*, 209–264.

Goss, A. K. (2007). Codifying a commons: Copyright, copyleft, and the Creative Commons project. *Chicago-Kent Law Review, 82*(2), 963–996.

Hatch, M. (2013). *The maker movement manifesto: Rules for innovation in the new world of crafters, hackers, and tinkerers.* McGraw-Hill Education.

Kennedy, D. M. (2001). A primer on open source licensing legal issues: Copyright, copyleft and copyfuture. *St. Louis University Public Law Review, 20*(2), 345–377.

Stallman, R. (1989). GNU General Public License. https://www.gnu.org/licenses/old-licenses/gpl-1.0.en.html

Suber, P. (2013). *Open access.* MIT Press. https://mitpress.mit.edu/books/open-access

Recommended Resources

Bailey, C. W. (2006). What is open access? In B. Pymm (Ed.), *Open access: Key strategic, technical and economic aspects* (pp. 13–26). Chandos Publishing.

Evans, J. A., & Reimer, J. (2009). Open access and global participation in science. *Science, 323*(5917), 1025. https://www.science.org/doi/abs/10.1126/science.1154562

Lewis, D. W. (2012). The inevitability of open access. *College & Research Libraries, 73*(5), 493–506. https://crl.acrl.org/index.php/crl/article/view/16255

Mallon, M. (2014). Maker mania. *Public Services Quarterly, 10*(2), 115–124. https://doi.org/10.1080/15228959.2014.904213

Mustonen, M. (2003). Copyleft—the economics of Linux and other open source software. *Information Economics and Policy, 15*(1), 99–121. https://doi.org/10.1016/S0167-6245(02)00090-2

Nascimento, S., & Pólvora, A. (2018). Maker cultures and the prospects for technological action. *Science and Engineering Ethics, 24*, 927–946. https://link.springer.com/article/10.1007/s11948-016-9796-8

See also: Creative Commons Licensing, Intellectual Property, Public Interest Design

66. Physical Computing

Emily F. Brooks

Physical computing pairs computer software and hardware to create an interactive experience. Physical computing generally refers to using a combination of sensors (input), microcontrollers (computation), and other material components (output). Sensors typically measure analog input from the environment, e.g., pressure, light, temperature, movement. The sensors are connected to a single-board microcontroller via digital and analog input/output pins. Microcontrollers are small (about 2x3 inches), inexpensive (about $22) computers that can perform basic calculations and input to output conversions, but have no peripherals like we might expect from a computer (monitor, mouse, keyboard, etc.). The data input from the sensors is then used to control things like motors, light-emitting diodes (LEDs), and piezos (vibrational sound elements).

219

Fig. 66.1. Digital circuitry designed with Meadow F7 Micro. Photo by Jorge Ramirez on Unsplash. https://bit.ly/keyword-fig-66–1

Arduinos are the most common microcontrollers used in humanities physical computing as they are open-source and the software is easy to use and well documented. The Arduino microcontrollers are useful for prototyping in a plug-and-play environment before creating more permanent projects with soldering and printed circuit boards (PCBs). Most parts can be purchased from online vendors, especially Arduino, Sparkfun, and Adafruit. Arduino also has a particularly useful language reference that explains the terminology in the code and provides examples of how it can be used.

An easy way to program the microcontroller is to connect it to a computer via USB and use the Arduino integrated development environment (IDE), which will verify the code, look for any bugs, and then upload it to the board, where it will be saved without requiring further connection to the computer. As long as the Arduino has a power supply, like a five-volt battery, it can continue to run the program in exhibits, field work, home installations, etc.

While physical computing projects are prevalent in STEM, they are also becoming popular in humanities pedagogy and research. At the University of Maryland, Kari Kraus assigned her students to augment the traditional book (as a genre), and student Clifford Hichar used the Lilypad Arduino to illuminate the illustrations with LEDs. A team at the University of Western Ontario used an Arduino to recreate a nineteenth-century card-playing magic trick automaton. A team at the University of Victoria also recreated an optophone using a Raspberry Pi (a single-board computer) and camera to take photos of texts and convert the images into sounds.

Physical computing is a major component of DIY and maker culture because of its experimental nature. The setup makes it easy to code, assess, and revise, often tweaking previously existing code to perform new actions desired. This also makes it a valuable learning experience for students in the classroom, to learn about the drafting and revision process. It is also a rewarding experience to see more immediate material outcomes of textual inputs, like making a light blink at one-second intervals with just seven lines of code. Some pedagogical implications of prototyping with physical computing are learning to communicate in a new language, identifying puzzles and seeking solutions, demystifying technology, and learning by making.

Recommended Resources

Adafruit. https://www.adafruit.com/
Arduino Language Reference. https://www.arduino.cc/reference/en/
Arduino Official Store. https://store.arduino.cc/usa/
Culkin, J. (2014). *Arduino!* [Informational comic] http://www.jodyculkin.com/wp-content/uploads/2014/03/arduino-comic-2014.pdf

Elliott, D., MacDougall, R., & Turkel, W. J. (2012). New old things: Fabrication, physical computing, and experiment in historical practice. *Canadian Journal of Communication, 37*(1). https://doi.org/10.22230/cjc.2012v37n1a2506

Hancock, C., Hichar, C., Holl-Jensen, C., Kraus, K., Mozafari, C., & Skutlin, K. (2013). Bibliocircuitry and the design of the alien everyday. *Textual Cultures, 8*(1). https://doi.org/10.14434/TCv8i1.5051

Igoe, T. (2008). *All about microcontrollers.* https://www.tigoe.com/pcomp/code/controllers/all-about-microcontrollers/

Igoe, T., & O'Sullivan, D. (2004). *Physical computing: Sensing and controlling the physical world with computers.* Thomson Course Technology.

Kinniburgh, M. C. (2017). *Arduino tutorial: Things to do in a maker space.* https://github.com/mckinniburgh/ArduinoTutorial/blob/master/ArduinoTutorial.md

Sayers, J. (2016). *Repository for physical computing and fabrication course at DHSI 2016.* https://github.com/uvicmakerlab/dhsi2016

Sparkfun. https://www.sparkfun.com/

Turkel, W. J., Sayers, J., & Elliott, D. (2014). *Physical computing and desktop fabrication for humanists.* http://dhsi.org/content/2014Curriculum/13.%20Physical%20Computing%20and%20Desktop%20Fabrication%20for%20Humanists%202014.pdf

See also: FabLab, Tinkering, Teardown, 3D Printing

67. Project Management

Sarah Young

The collaborative and iterative nature of maker projects leads to a growing focus on project management proficiency in and outside of educational contexts. Project management is a necessary workplace skill increasingly important to employers and educational programs (Berggreen & Kampf, 2015; Miller & Hovde, 2018; Pope-Ruark, 2015; Whiteside, 2003). According to the Project Management Institute (PMI) (2008), a project is a temporary endeavor creating "a unique product, service, or result" with a "definite beginning and end" that concludes when the project is complete, when it is determined the objectives won't be met, or when the project is no longer required.

A project is unique in that although some of the steps to create it may be institutionally repeated, each project has its own distinct needs. A project is temporary because although the project deliverable itself is usually designed with permanence, once the end conditions are met, the project is over. For instance, one team may work on constructing many buildings, but each building site is a project and would have unique expectations according to location. Further, after the building is built, although the processes to construct a building may be momentary, the building's tenure may be indefinite. Moreover, PMI continues that project management is "the application of knowledge, skills, tools, and techniques to project activities to meet the project requirements" (2008, p. 6).

This involves the five processes: initiating, planning, executing, monitoring and controlling, and closing. The processes are carried out through steps like identifying requirements, addressing stakeholder expectations, and overcoming constraints (among many, like quality, budget, and time). Each step is completed in association with the other steps and each provides structure to accomplish project goals.

Because projects span various organizations, social influences, and cultures, an important part of project management is (various types of) communication (Berggreen & Kampf, 2015; Miller & Hovde, 2018; Pope-Ruark, 2015; Whiteside, 2003). First, project team members need basic interpersonal skills like how to interact with others (Pope-Ruark, 2015). Second, project members also need more specific organizational and cultural skills to work with other members of the project team through workplace skills like engaging in open discussions, recognition of good team behavior, comprehension of team communication rules like how to participate in meetings and resolve conflict, and the ability to articulate project tasks (Pope-Ruark, 2015). Members must also possess organizational and directional skills to manage (multiple) projects and team members, and it is important to understand corporate cultures like "politics and style of business" (Whiteside, 2003, p. 311). Third, team members must gather, write, and present information clearly not just between each other but also to invested parties outside the project team that may be unfamiliar with a team's workplace culture or the day-to-day breakdown of the project (Miller & Hovde, 2018). All these skills are also in the backdrop of technology. Whether project team members are working in a technological milieu like engineering (Whiteside, 2003), or communicating through technology (Miller & Hovde, 2018; Whiteside, 2003), sociotechnical skills are essential in understanding context, content, and delivery of messages (Berggreen & Kampf, 2015).

Overall, while there are many moving processes involved in project management, successful completion of goals hinges on completing a series of procedures and managing communication between stakeholders both inside and outside a project team with a rhetorical awareness of project expectations, organizational cultures, and technology.

223

References

Berggreen, L., & Kampf, C. (2015). Project management communication 2.0 - The socio-technical design of PM for professional communicators. *2015 IEEE International Professional Communication Conference* (pp. 1–9). IEEE.

Miller, R. T., & Hovde, M. R. (2018). Technical communication and project management: A mixed methods study in a corporate context. *2018 IEEE International Professional Communication Conference* (pp. 173–180). IEEE.

Pope-Ruark, R. (2015). Introducing agile project management strategies in technical and professional communication courses. *Journal of Business and Technical Communication, 29*(1), 112–133.

Project Management Institute. (2008). *A guide to the project management body of knowledge (PMBOK guide)* (4th ed). Project Management Institute, Inc.

Whiteside, A. (2003). The skills that technical communicators need: An investigation of technical communication graduates, managers, and curricula. *Journal of Technical Writing and Communication, 33*(4), 303–318.

Recommended Resources

Bergmann, T., & Karwowski, W. (2019). Agile project management and project success: A literature review. In *Advances in Human Factors, Business Management and Society: Proceedings of the AHFE 2018 International Conference on Human Factors, Business Management and Society* (pp. 405–414). Springer International Publishing.

Ivetić, P., & Ilić, J. (2020). Reinventing universities: Agile project management in higher education. *European Project Management Journal, 10,* 64–68. https://www.ceeol.com/search/article-detail?id=924582

Kerzner, H. (2022). *Project management metrics, KPIs, and dashboards: A guide to measuring and monitoring project performance.* John Wiley & Sons.

Lock, D. (2020). *Project management* (10th ed.). Routledge.

Pollack, J. (2007). The changing paradigms of project management. *International Journal of Project Management, 25*(3), 266–274. https://doi.org/10.1016/j.ijproman.2006.08.002

See also: Technical Communication, DIWO, Shared Leadership

68. Public Interest Design

Max Renner

In 1968, American Civil Rights leader Whitney Young issued a chal-lenge to attendees of the American Institute of Architects (AIA) nation-al convention:

> ". . . you are not a profession that has distinguished itself by your so-cial and civic contributions to the cause of civil rights, and I am sure this does not come to you as any shock. You are most distinguished by your thunderous silence and your complete irrelevance."

Although this rebuke from Young could be perceived as a rather harsh cri-tique, it is hard to ignore the fact that architecture, as Elizabeth Grosz (2001) has noted, is "probably the largest, most systematic and most powerful mode of spatial organization and modification" (p. 110). As such, architecture is always implicated in our lives and architects of the time were being taken to task for how they understood their role as ethical practitioners. Although a much more recent development, it is from Young's call and the 1968 AIA convention, that public interest design traces its roots.

225

Fig. 68.1. Subway systems and spaces are examples of public-facing design that need special attention to public interests. Photo by Viktor Forgacs on Unsplash. https://bit.ly/keyword-fig-68–1

Public interest design is a movement within the architectural profession that calls for a reconceptualization of the relationship of architects and designers to their projects. The origins and commitments of public interest design emerged from a critique of the architectural profession regarding ethics. Public interest design is predicated on the notion that architects should be designing for communities rather than for clients (Abendroth & Bell, 2016). The traditional model for architects has been to design and work for a client, whether that be an organization, family, or individual. Public interest design posits that architects have a responsibility to work with communities more broadly. This intervention into architecture necessitates a reconsideration of not only the process of working on a project, but also the relationship of an architect to a project.

Designing for a community, rather than a client, entails engaging people well beyond those an architect may be able to easily access; it also recognizes that the work of architects does far more and impacts more lives than simply those who may fund their projects.

References

Abendroth, L. M., & Bell, B. (Eds.). (2015). *Public interest design practice guidebook: SEED methodology, case studies, and critical issues*. Routledge.

The American Institute of Architects. (1968). Full Remarks of Whitney M. Young Jr. AIA Annual Convention in Portland, Oregon June 1968. http://content.aia.org/sites/default/files/2018–04/WhitneyYoungJr_1968AIAContention_FulLSpeech.pdf

Grosz, E. (2001). *Architecture from the outside: Essays on virtual and real space*. MIT Press.

Recommended Resources

El-Zein, A. H., & Hedemann, C. (2016). Beyond problem solving: Engineering and the public good in the 21st century. *Journal of Cleaner Production, 137*, 692–700. https://doi.org/10.1016/j.jclepro.2016.07.129

Golsteijn, C., Gallacher, S., Capra, L., & Rogers, Y. (2016, June). SensUs: Designing innovative civic technology for the public good. In *Proceedings of the 2016 ACM Conference on Designing Interactive Systems* (pp. 39–49). ACM. https://dl.acm.org/doi/abs/10.1145/2901790.2901877

Junginger, S. (2017). Design research and practice for the public good: A reflection. *She Ji: The Journal of Design, Economics, and Innovation, 3*(4), 290–302. https://doi.org/10.1016/j.sheji.2018.02.005

Niemiec, R., McCaffrey, S., & Jones, M. (2020). Clarifying the degree and type of public good collective action problem posed by natural resource management challenges. *Ecology and Society*, 25(1), Article 30. https://doi.org/10.5751/ES-11483-250130

UK Design Council. (2013). Design for public good. *Annual Review of Policy Design*, 1(1), 1–50.

See also: Social Innovation, Inclusive Design, Advocacy, Accessibility

69. Radical Collaboration

Joe Moses

H ere's how the Stanford d.school—the institution responsible for pop-ularizing design thinking as a mode of collaborative problem solving—defines radical collaboration:

> To inspire creative thinking, we bring together students, faculty, and practitioners from all disciplines, perspectives, and backgrounds—when we say radical, we mean it! Different points of view are key in pushing students to advance their own design practice. By 'radical' in Stanford's definition, they're talking about authentically *diverse*. When we talk about *collaboration,* we're talking about people with diverse interests and experiences who pursue a common goal. (n.d.)

Radical Collaboration in Practice. A team at the University of Minnesota used radical collaboration while researching emerging technologies—for example new wearable devices such as Google Glass and Apple Watch—from the perspectives of undergraduate students, graduate students, instructors, and senior tenured faculty. The team set out on a collaborative research and writing project that included findings from everyone on the research team. Together they explored the impact of wearable technologies in teaching and industry. By including perspectives of researchers across borders of age and academic rank, and by using methods suited to researchers across borders of academic experience, the team successfully produced an article for publication in *Connexions: International Professional Communication Journal* (Duin et al., 2017).

What's radical about this example of collaboration is that it differs significantly from traditional academic practices (Duin, Moses, McGrath, Tham, & Ernst, 2017, p. 65) with respect to leadership, scaffolding, direction, and problem-solving:

- Share leadership, research, and teaching roles (not top-down; not bottom-up)
- Expose participants to the complexities of problems regardless of experience (not scaffolded)
- Resist hierarchical structures (used direction from all participants; acted on direction from all participants)
- Suspend closure: sustain problem definition (spent as much or more time defining the problem as proposing a solution)

As the example suggests, what's radical about radical collaboration is the suspension of traditional practice, which leads to traditional outcomes, for radical practices that lead to radical outcomes (that is, outcomes that work but that no one expects). Robert Verganti (2009) in his book, *Design-Driven Innovation: Changing the Rules of Competition by Radically Innovating What Things Mean,* suggests that traditional practices we take for granted and assume to be foundational to sound methods may get in the way of the goals we pursue when we adopt them. For example, user-centered design—whose fundamental premise is that the best way to solve human problems is to talk to the people who experience them—can get in the way of innovation because of individuals' limited experience and imaginations (p. 49). Radical collaboration, to use Verganti's words, is a *proposal,* and a *vision* (p. 10) for a solution that no one may even have asked for.

Fig. 69.1. In one of our research team meetings, we did an exercise where everyone, no matter rank nor role, wrote their desired research objectives for the team as a way to determine the direction for the year. A short time-lapse video can be accessed here: https://fb.watch/keYL-Vo7lp/. Video and screen capture provided by Jason Tham.

Radical is as Radical Does. What people who embark on radical collaboration are asking for is a test of their assumptions about 1) what collaboration is, 2) how collaboration works, and 3) how to deal with vulnerabilities that arise in the collaborative process. The key word of James W. Tamm and Ronald J. Luyet's book *Radical Collaboration: Five Essential Skills to Overcome Defensiveness and Build Successful Relationships* is "defensiveness"—the one renewable resource among human collaborators that must be depleted before productive collaboration can take place. To deplete defensiveness, the authors outline five skills for collaborators—achieving an attitude of collaboration, of honesty, of responsibility, of awareness, and of negotiated problem solving—all of which are among the "soft skills" that organizations of all sizes have begun to add to their job descriptions.

When collaboration becomes ordinary enough that researchers and organizations can define what effective collaboration is, the theory of radical collaboration says that whatever we know about collaboration pales next to what we haven't yet thought to learn about it.

References

Duin, A. H., Moses, J., McGrath, M., Tham, J., Ernst, N. (2017). Design thinking methodology: A case study of "radical collaboration" in the wearables research collaboratory. *Connexions: International Professional Communication Journal 5*(1), 45–74. https://connexionsj.files.wordpress.com/2017/06/duin-etal1.pdf

Tamm, J. W., & Luyet, R. J. (2004). *Radical collaboration: Five essential skills to overcome defensiveness and build successful relationships.* Harper Collins.

Verganti, R. (2009). *Design-driven Innovation: Changing the rules of competition by radically innovating what things mean.* Harvard Business School Publishing.

Recommended Resources

Bene, R., & McNeilly, E. (2020). Getting radical: Using design thinking to tackle collaboration issues. *Papers on Postsecondary Learning and Teaching, 4,* 50–57. DOI: https://doi.org/10.11575/pplt.v4i.68832

Duin, A. H., Tham, J., & Moses, J. (2019). Connecting silos through design thinking: Radical collaboration in the Wearables Research Collaboratory. In P. Minacori (Ed.), *Technical communication: Beyond the silos* (pp. 51–65). Hermann Publications.

Hamm, S. (2008). Radical collaboration: Lessons from IBM's innovation factory. *Human Resource Management International Digest, 16*(2). https://doi.org/10.1108/hrmid.2008.04416bad.009

Kukla, R. (2012). "Author TBD": Radical collaboration in contemporary biomedical research. *Philosophy of Science, 79*(5), 845–858. DOI: https://doi.org/10.1086/668042

Simons, T., Gupta, A., & Buchanan, M. (2011). Innovation in R&D: Using design thinking to develop new models of inventiveness, productivity and collaboration. *Journal of Commercial Biotechnology, 17,* 301–307. https://doi.org/10.1057/jcb.2011.25

See also: Radical Imagination, Project Management, Shared Leadership

70. Safety

R.J. Lambert

When people refer to safety, they often mean *personal safety* (Chmiel & Grote, 2017, p. 399). According to Abraham Maslow's Hierarchy of Needs, the psychological need for safety is second only to fundamental physiological necessities, such as food, hydration, and sleep. Personal safety could be described as the absence of harm or danger. Since people perceive harm and danger differently, safety may also be understood as a person's subjective experience relative to harm or danger. This is the difference between *being* safe and *feeling* safe. Personal safety reflects a person's fundamental desire for stability (Poston, 2009, p. 350).

Since safety often depends on the stable function of instruments and processes, the term also refers to a lack of mechanical or structural failure(s). This second kind of safety is called *process safety* (Chmiel & Grote, 2017, p. 399) or *reliability* (Nathwani et al., 2006). In other words, a safe instrument or process is one in which failure has been prevented, minimized, or otherwise addressed. Carl Forssell's (1924) "principle of optimality" says that design should minimize total costs, which include production costs as well as the expected costs of failure (Elishakoff, 2004, p. 272; Nathwani et al., 2006, p. 265).

In engineering, the safety or reliability of designs is predicted by the *safety factor*, or maximum load, of each material and part (Madsen et al., 2006). The difference between the maximum possible load and the actual (expected) load is called the *safety margin*, which approximates how far a design is from risking failure. Safety factors and safety margins are documented in a *safety index*, which can guide design to increase overall safety (Madsen et al., 2006).

Design thinking foregrounds users' needs and desires in order to minimize failures and increase both process safety and personal safety. In order to address design failures, the causes may be identified and documented with *safety checks* and *usability testing* throughout design and production.

Moving outside of personal safety and process safety, the terms *safety climate* and *safety culture* describe broad contexts that contribute to and/or result from safety (Chmiel & Grote, 2017, p. 404). E. Scott Geller (2016) describes how changes in infrastructure, attitude, and behavior may achieve "the ultimate vision of a safety improvement mission": a Total Safety Cul-

231

ture. The coordination of optimized safety systems, at the institutional level, and workers' active safety precautions, at the individual level, contributes to a Total Safety Culture where the value of safety is linked to all situation priorities (Geller, 2016).

Healthcare is one industry that consistently implements safety measures at all levels. Given the high cost of failure in healthcare delivery, hospitals create a safety culture that reduces medical errors and promotes patient health. An example of design thinking in healthcare delivery is the development of haptic feedback devices to reduce syringe swap and manual injection errors in anesthesiology. Researchers designed six prototypes which attached to existing syringes, and volunteers tested for the adequate level of haptic feedback and resistance to prevent syringe errors (Williams et al., 2019). In the broader hospital context, nursing research suggests that improving care transitions, staffing adequately, and changing how errors are handled may contribute to an organizational culture of safety (Hershey, 2015).

References

Chmiel, N., & Grote, G. (2017). Why do I put myself and others in danger or help increase safety? Person- and situation-related causes of safety behaviours. In N. Chmiel, F. Fraccaroli, & M. Sverke (Eds.), *An introduction to work and organizational psychology: An international perspective* (pp. 388–409). Wiley Blackwell.

Elishakoff, I. (2004). *Safety factors and reliability: Friends or foes?* Kluwer Academic Publishers.

Forssell, C. (1924). Ekonomi och Byggnadsvasen, Sunt Fornoft. English translation in N. C. Lind (1970), *Structural reliability and codified design* (pp. 74–77). Waterloo University Press.

Geller, E. S. (2016). *The psychology of safety handbook* (2nd ed.). CRC Press.

Hershey, K. (2015). Culture of safety. *Nursing Clinics of North America, 50*(1), 139–152. https://doi.org/10.1016/j.cnur.2014.10.011

Madsen, H. O., Krenk, S., & Lind, N. C. (2006). *Methods of structural safety*. Dover.

Nathwani, J. S., Mahesh, D. P., & Lind, N. S. (2006). A standard determination of optimal safety in engineering practice. In J. D. Sørensen and D. M. Frangopol (Eds.), *Advances in reliability and optimization of structural systems* (pp. 261–268). CRC Press.

Poston, B. (2009). An exercise in personal exploration: Maslow's hierarchy of needs. *The Surgical Technologist, 347*–353. http://www.ast.org/pdf/308.pdf

Williams, D., Eagle, B., & Dingley, J. (2019). Development of a haptic feedback device to reduce syringe substitution and drug overdosage error. *Anaesthesia*, 74(11), 1416–1424. https://doi.org/10.1111/anae.14736

Recommended Resources

Bharti, N. (2017). 3D printing in makerspaces: Health and safety concerns. *Issues in Science and Technology Librarianship, (87)*. https://doi.org/10.29173/istl1712

Cooper, D. (2003). Psychology, risk and safety. *Professional Safety*, 48(11), 39–46. https://www.researchgate.net/publication/285258014_Psychology_risk_and_safety

Love, T. S., Roy, K. R., & Marino, M. T. (2020). Inclusive makerspaces, fab labs, and STEM labs. *Technology and Engineering Teacher*, 79(5), 23–27.

Spencer, T., Spencer, V., Patel, P., & Jariwala, A. (2016, November). Safety in a student-run makerspace via peer-to-peer adaptive training. In *Proceedings of ISAM 2016: International Symposium on Academic Makerspaces* (pp. 1–6).

Stark, J., Anderson, F., Fitzmaurice, G., & Somanath, S. (2020, July). Makeaware: Designing to support situation awareness in makerspaces. In *Proceedings of the 2020 ACM Designing Interactive Systems Conference* (pp. 1005–1016).

233

See also: Cognitive Dissonance, Embodiment, Extreme Situation, Tools

71. Shared Leadership

Ann Hill Duin

Shared leadership has many names, including partnership as leadership, distributed leadership, and community of leaders. Successful shared leadership requires a balance of power, shared purpose and goals, shared responsibility for the work, respect for each person, and a willingness to work together closely on complex, real-world needs. Maker culture, maker education, makerspaces, and design thinking all entail components of shared leadership.

The scholarly community's understanding of shared leadership has evolved, beginning with a focus on looking beyond the designated leader for guidance. In 1924, Mary Parker Follett introduced the *law of the situation*, i.e., that conflicts should be resolved according to the facts of the situation and not by reference to the relative superiority of any party over the other. Throughout her writing she emphasized that "The best leader does not ask people to serve him, but the common end. The best leader has not followers, but men and women working with him" (Stoner, 2019, n.p.). Today, the emphasis is on engagement and mutual influence to achieve shared goals. Craig L. Pearce, Charles C. Manz, and Henry P. Sims (2008) define shared leadership as "a process where all members of a team are fully engaged in the leadership of the team: Shared leadership entails a simultaneous, ongoing, mutual influence process involving the serial emergence of official as well as unofficial leaders" (p. 353).

International researchers Jürgen Weibler and Sigrid Rohn-Endres (2010) further define *shared network leadership* as "networks of individuals engaged in reciprocal, preferential, mutually supportive actions . . . [in which the individuals] agree to forego the right to pursue their own interests at the expense of others" (p. 182). Moreover, "Leadership emerging from the collective is embedded in a certain quality of network relationships and requires a certain learning environment" (p. 186). Based on their study of interorganizational networks, Weibler and Rohn-Endres (2010) emphasize the evolutionary change in dialogue and relationships that occurs as shared network leadership develops (Fig. 71.1).

- In stage one, collaborators are polite: they largely repeat rules, reproduce rules and existing knowledge (mainly from their individual perspectives), and there is little responsibility for joint tasks.

· In stage two, collaborators are more authentic, beginning to reveal how they see and interpret conflicts.

· At stage three, participants "must leave blaming speech act behind and cultivate self-reflection and empathetic listening . . . The actors start carrying out this shift by cultivating a stance of inquiry that is rooted in a spirit of curiosity." Collaborators give voice to differences and are interested "in the multitude of other views, in other persons and how they deal with their problems" (p. 191).

· At stage four, participants "speak in the 'we' form, refer to the emotionally supportive character of the network based on common experience, [and] emphasize their common ground" (p. 191).

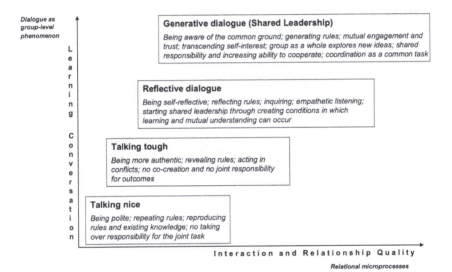

Fig. 71.1. Development of a learning conversation and emergence of shared leadership (Weibler & Rohn-Endres, 2010).

A comparison of the characteristics of vertical, shared, and shared network leadership is useful as one works to identify and foster a culture of shared leadership (Table 70.1). For example, to identify shared leadership in action, watch for conversations that flow in rhythm to those possessing the most relevant knowledge to offer regarding the problem or opportunity of the moment. The most needed input should come from those best able to contribute. Again, makerspaces inherently promote a culture of shared leadership through enhancing the ability to identify knowledge sets, providing support for interdependence and multiple means for enhancing process, and eliciting constant communication.

Table 70.1. A comparison of the characteristics of vertical, shared, and shared network leadership.

Vertical Leadership	Shared Leadership	Shared Network Leadership
Identified by position in a hierarchy and downward influence from a superior	Identified by individuals' knowledge sets and consequent abilities to influence peers	Identified by outcomes of interorganizational networks
Evaluated by whether the leader solves problems or not	Evaluated by how well people are working together	Evaluated by how well people give voice to differences and emotionally support the whole
Leaders provide solutions and answers	Leaders provide multiple means to enhance process	Group as a whole shares responsibility on common tasks
Distinct differences between leaders and followers	Members are interdependent	Members transcend self-interest
Communication is formal	Communication is critical	Communication is generative

I contend that the future of higher education is about designing the conditions to foster shared leadership and shared network leadership. Maker culture, maker education, makerspaces, and design thinking represent the conditions that best foster shared leadership. In makerspaces, the group as a whole can share responsibility on common tasks, members can transcend self-interest, and communication is generative. Such a mindset and practice of collaborating with courage amid recognition of culture provides a clear leverage for accomplishing targeted goals and adding value to the overall work underway. This is the challenge before us: to promote and foster makerspaces and shared leadership as a means to improve student success and foster the future sustainability of higher education.

References

Follett, M. P. (1924) *Creative experience*. Longman Green & Co. (Reprinted by Peter Owen in 1951).

Pearce, C. L., Manz, C. C., & Sims, H. P., Jr. (2008). The roles of vertical and shared leadership in the enactment of executive corruption: Implications for research and practice. *The Leadership Quarterly, 19,* 353–359.

Pearce, C. L., Manz, C. C., & Sims, H. P., Jr. (2014). *Share, don't take the lead*. Information Age.

Stoner, J. L. (2019). 15 quotes by Mary Parker Follett — Guidance for today's world. https://seapointcenter.com/15-quotes-by-mary-parker-follett/

Weibler, J., & Rohn-Endres, S. (2010). Learning conversation and shared network leadership. *Journal of Personnel Psychology, 9*(4), 181–194.

Recommended Resources

Bolden, R., Jones, S., Davis, H., & Gentle, P. (2016). *Developing and sustaining shared leadership in higher education*. Leadership Foundation for Higher Education.

Duin, A. H., & Baer, L. L. (2010). Shared leadership for a green, global, and Google world. *Planning for Higher Education*, July-Sept issue, 30–38.

Duin, A. H., Cawley, S., Gulachek, B., O'Sullivan, D., & Wollner, D. (2011). Shared leadership transforms higher education IT. *EDUCAUSE Quarterly, 34*(2), April-June.

Pearce, C. L., & Conger, J. A. (2003). *Shared leadership: Reframing the hows and whys of leadership*. Sage Publications.

Ulhoi, J. P., & Muller, S. (2014). Mapping the landscape of shared leadership: A review and synthesis. *International Journal of Leadership Studies, 8*(2), 66–87.

See also: Radical Collaboration, Radical Imagination, DIWO

72. Social Innovation

Nupoor Ranade

Social innovation is a process through which collaboration takes place between individuals and organizations in order to solve problems that affect civic life. More often than not, the motivation for social innovation projects comes from social needs (Mulgan, 2006) like economic conditions, education, health and well-being, etc. The word *innovation* is used in a broad sense covering the use of knowledge for creation and introduction of something "new" (Holt, 1971). In this case, it refers to the development of a new idea, even if it is new only to the actors participating in the process. Innovation is also concerned with change. Social innovation is targeted towards bringing about an improvement in current conditions in order to meet social goals.

Social innovation is relevant in every sector but is more prominent in fields like activism (diversity and feminism movements), online crowdsourcing (Wikipedia and open-source applications), microcredit (funding for those who lack collateral, steady employment, or a verifiable credit history) and distance learning (The Open University and models for distance education). These movements have never been restricted to the nonprofit sector. Collaborations between the government, markets, academia as well as social enterprises are on the rise. The decisive role of users in social innovation is increasingly being recognized in private businesses too.

What we now count as progress has come about through the mutual reinforcement of social, economic, technological and political social innovations led by social movements such as anti-slavery, trade unions, cooperatives or even reading clubs. The history of social innovation dates back to the late eighteenth century. The anti-slavery movement pioneered the use of social capital for methods like campaigns, demonstrations, petitions, consumer boycotts, logos and slogans (Mulgan et. al., 2007). Research published by social innovators like Edwin Chadwick in the nineteenth century, persuaded the government to provide clean water, sewers and street cleaning (Mulgan et. al., 2007). The twentieth century witnessed a rise in social movements around ecology, feminism and civil rights which spawned innovations in governments and commercial markets. Religion played a role in creating and sustaining social innovation (Mulgan et. al., 2007). For example, the Muslim NGO called Islamic Relief is engaged in responding to emergencies & fighting poverty. Similarly, US Black churches were instrumental in the civil rights movement and innovations in micro-banking (Mulgan et. al., 2007). Due to the multitude of solutions offered by social innovation, the term frequently gets replaced by terms such as social entrepreneurship, corporate social responsibility, and social enterprise.

What is the relationship between social innovation and social entrepreneurship? The dichotomy lies in the motivation for the two sectors. Social entrepreneurships are enterprises with a social betterment goal that are structured to make a profit (Foreman, n.d). For example, DlightDesign is a social entrepreneurship that sells solar lights to communities that don't have reliable electricity. The company has to depend on its own profits for future expansion, like a start-up. Social innovation, on the other hand, questions the premises on which existing social structures are built and then reimagines systems and institutional relationships to bring about change. An example is fair trade, whereby all parties are involved in deriving some benefit, but the overarching emphasis is on creating social value and benefiting society as a whole rather than gains for private individuals (Phills et al., 2008). This motivation of social innovation also differentiates it from business innovations, which are generally based on profit maximization over civic benefits.

239

A literature review of social innovation reveals variations of its definition by not only different researchers but also in different fields including public policy marketing, technology management, culture and organization, business research, organization science, and so on (Agostini et. al., 2017). Useful attempts are being made to understand social innovation at university levels. Stanford publishes an online publication called *Stanford Social Innovation Review*, Duke's Fuqua School of Business has done significant work on social entrepreneurship and Harvard runs an extensive program on innovations in governance. In the UK, research on social innovation is being conducted through the lens of public administration and business. With the advent in technology, the nature of activities carried out for social innovation is going through a drastic change. A subdomain of social innovation has been defined in relation to the introduction of digital technologies. Sussex and Manchester are researching social innovation from a technological perspective that includes digital social innovation. A digital innovation online project called DSI4EU is being carried out by a consortium of seven partner organizations: Nesta (UK), Waag (Netherlands), betterplace lab (Germany), Fab Lab Barcelona (Spain), WeMake (Italy), Barcelona Activa (Spain) and the ePa stwo Foundation (Poland).

Geoff Mulgan quotes Lord Macauley's prose: "There is constant improvement precisely because there is constant discontent" (Mulgan, 2006,

p. 148). Social innovation is the best construct for understanding-and pro-
ducing-lasting social change (Phills, 2008). In order to create sustainable
social innovations, it is important to utilize research happening in the community while also embracing interdisciplinary cooperations (Taylor, 1970).

References

About the project - DSI4EU. (n.d.). https://digitalsocial.eu/
about-the-project

Agostini, M. R., Vieira, L. M., Tondolo, R. (2017). An overview on social
innovation research: Guiding future studies. *Brazilian Business Review,*
14(4), 385–402.

Foreman, M. (n.d.). Social innovation and entrepreneurship. https://www.dickinson.edu/homepage/687/
social_innovation_and_entrepreneurship

Holt, K. (1971). Social innovations in organizations. *International Studies of*
Management & Organization, 1(3), 235–252.

Mulgan, G. (2006). The process of social innovation. *Innovations: Technology,*
Governance, Globalization, 1, 145–162.

Mulgan, G., Tucker, S., Ali, R., Sanders, B., University of Oxford, &
Skoll Centre for Social Entrepreneurship. (2007). *Social innovation: What*
it is, why it matters and how it can be accelerated. Young Foundation.

Phillips, W., Lee, H., Ghobadian, A., O'Regan, N., & James, P. (2015).
Social innovation and social entrepreneurship: A systematic review.
Group & Organization Management, 40(3), 428–461.

Phills, J. A., Deiglmeier, K., & Miller, D. T. (2008). Rediscovering social
innovation. *Stanford Social Innovation Review, 6*(4), 34–43.

Taylor, J. B. (1970). Introducing social innovation. *The Journal of Applied Be-*
havioral Science, 6(1), 69–77.

Recommended Resources

Howaldt, J., Domanski, D., & Kaletka, C. (2016). Social innovation: Towards a new innovation paradigm. *RAM. Revista de Administração Mackenzie,*
17, 20–44.

Manzini, E. (2015). *Design, when everybody designs: An introduction to design for social*
innovation. MIT Press.

Nicholls, A., & Murdock, A. (2012). The nature of social innovation.
In A. Nicholls and A. Murdock (Eds.), *Social innovation: Blurring bound-*
aries to reconfigure markets (pp.1–30). Palgrave Macmillan. https://doi.
org/10.1057/9780230367098_1

Osburg, T., & Schmidpeter, R. (Eds.). (2013). *Social innovation: Solutions for a*
sustainable future. Springer.

Wittmayer, J. M., Backhaus, J., Avelino, F., Pel, B., Strasser, T., Kunze, I., & Zuijderwijk, L. (2019). Narratives of change: How social innovation initiatives construct societal transformation. *Futures, 112*, 102433. https://doi.org/10.1016/j.futures.2019.06.005

See also: Inclusive Design, Public Interest Design, Community Engagement

73. STEM & STEAM

Mary E. Caulfield

In response to an increasing need to equip students for careers in fields requiring quantitative and scientific aptitude, both K-12 and college-level education have focused on improving the quality of education in these subject areas. The term STEM (Science, Technology, Engineering, and Mathematics) originally referred to increasing students' competence in these individual subjects, but has increasingly been used to refer to the interdependence between the fields. More recently, in response to the need for students to develop skills in design, critical thinking, and hands-on work, the letter A (for Arts) has been added, and the term STEAM has come into frequent use. The emphasis on improving rigor and competency in these fields and in understanding the interplay between design, science, and quantitative skills has led educators to consider how this interdependency can be implemented and assessed.

As the emphasis has changed from competence in individual subject areas to facility with a constellation of skills, STEM/STEAM has come to be associated with a wide range of initiatives ranging from teaching math or science together with one or more of the arts to projects demanding multiple skills in areas such as technology, science, and communications. This multidisciplinary approach is a response to trends in industry and research that require understanding of the ways in which formerly distinct fields, such as computer science and biology, now interact. The addition of arts education shows the potential of the arts to aid in design thinking. For example, the use of origami in studying structures reflects the intersection of arts, mechanical engineering, and materials science.

While this holistic approach has much to show both educators and students about the ways that different disciplines can work together, concerns have been raised about the need for mastery of distinct disciplinary concepts as well as the need for educators to have sufficient training in the quantitative and scientific fields. This level of teacher training is particularly important at the K-12 level, where teacher education emphasizes child development and less emphasis may be placed on subject area knowledge. Project-based learning has been seen as one approach for an integrated STEM/STEAM curriculum, but concerns have been raised about whether this approach can teach discipline-specific knowledge to primary and elementary-level learners. In addition, while many opportunities exist for professionals from different fields to work together, there is still a need for secondary and bachelors-level students to know their fields in depth and differentiate between ways of thinking that are unique to each discipline.

To implement a STEM/STEAM approach within cohesive and research-based curricula, educators across all levels will need to work together to address academic needs in ways that are both rigorous and developmentally appropriate. In addition, inequalities between school systems with widely divergent resources will need to be addressed. While innovative programs have shown the possibilities that exist for challenging motivated students and engaging them intellectually, a case will need to be made for providing additional resources and teacher training, as well as for cooperation among educators from the primary to college levels.

Recommended Resources

Bequette, J. W., & Bequette, M. B. (2012). A place for art and design education in the STEM conversation, *Art Education, 65*(2) 40–47.

Guyotte, K. W., Sochacka, N. W., Costantino, T. E., Kellam, N., Kellam, N. N., & Walther, J. (2015). Collaborative creativity in STEAM: Narratives of art education students' experiences in transdisciplinary spaces. *International Journal of Education & the Arts, 16*(15). http://www.ijea.org/v16n15/

National Research Council (2014). *STEM integration in K–12 education: Status, prospects, and an agenda for research*. The National Academies Press. https://doi.org/10.17226/18612

Resnick, M. (2017) *Lifelong kindergarten: Cultivating creativity through projects, passion, peers, and play*. MIT Press.

Stohlmann, M., Moore, T., & Roehrig, G. (2012). Considerations for teaching integrated STEM education. *Journal of Pre-College Engineering Education Research, 2*(1), Article 4. https://docs.lib.purdue.edu/cgi/viewcontent.cgi?article=1054&context=jpeer

See also: Active Learning, Maker Competencies, Technical Communication

74. Technical Communication

Sara Doan

Technical communication is about writing and designing content that aims to enable people, or users, to accomplish tasks across workplace, professional, and civic contexts. In this definition, technical means task- or purpose-based communication. Examples of technical communication include user manuals, help guides, computer notifications, technical descriptions, and instances where technical or procedural information needs to be delivered to users. Technical communicators, or those who create technical communication, "make information more usable and accessible to those who need that information, and in doing so, they advance the goals of the companies or organizations that employ them" (Society for Technical Communication, n.d.).

Beyond examples of communication artifacts, technical communication exists as both an academic and applied discipline, or way of developing "knowledge in producing complex information for a wide variety of audiences" (Melonçon, 2018, p. 202). Technical communication became a professional discipline during the late nineteenth and early twentieth centuries (Russell, 2004). Between the 1940s and 2000, technical communicators largely worked with engineers to develop and document new technologies (Malone, 2011). Around 2000, technical communication affiliated itself with Information Technology, expanding into the digital age with usability, user experience, and user-centered design (Dicks, 2010).

A more in-depth example of technical communication is a user manual for assembling an IKEA bookcase. A user manual is a written or picture-based set of instructions that a person, or user, employs for direction about how to complete a task or set of tasks. When users want to assemble a new bookcase, they consult the steps outlined in the user manual. Communicating warnings about safety risks is central to technical communication, such as including information about how to secure a tall bookcase to an adjoining wall in order to prevent the bookcase from tipping and falling onto a small child. Along with written and visual instructions, technical communication occurs digitally. For example, in office enterprise software like Microsoft Outlook or Google Docs, clicking the "Help" button will take the user to an online manual. Although technical communication is generally defined as workplace writing, instances of technical communication regularly appear outside of organizational contexts; for example, Reddit's "Explain Like I'm Five" subreddit contains technical descriptions of ordinary events and objects (Pflugfelder, 2017).

However, design thinking and usability complicate both these examples and the practice of technical communication. While instructional genres like user manuals or digital troubleshooting guides and skills in plain language, visual design, and project management are important to technical communication, the most important function of technical communication is advocating for users during product or digital design processes (Hart-Davidson, 2013). Technical communicators must consider how their users engage with technologies even without user manuals. Recent products, such as virtual reality headsets and smartwatches, have been created to be used intuitively, not requiring users to read a manual before using a product for the first time. Quality technical communication requires understanding how users engage with technologies and instructions in order to create digital, visual, aural, or written texts enabling users to complete tasks safely.

References

Dicks, R. S. (2010). The effects of digital literacy on the nature of technical communication work. In R. Spilka (Ed.), *Digital literacy for technical communication: 21st century theory and practice* (pp. 51–81). Routledge.

Hart-Davidson, W. (2013). What are the work patterns of technical communication? In J. Johnson-Eilola and S.A. Selber (Eds.), *Solving problems in technical communication* (pp. 50–73). University of Chicago Press.

Malone, E. (2011). The first wave (1953–1961) of the professionalization movement in technical communication. *Technical Communication, 58*(4), 285–306.

Meloçon, L. K. (2018). Critical postscript: On the future of the service course in technical and professional communication. *Programmatic Perspectives, 10*(1), 202–230.

Pflugfelder, E. H. (2017). Reddit's "Explain like I'm five": Technical descriptions in the wild. *Technical Communication Quarterly, 26*(1), 25–41. https://doi.org/10.1080/10572252.2016.1257741

Russell, D. R. (2004). The ethics of teaching ethics in professional communication: The case of engineering publicity at MIT in the 1920s. In J. M. Dubinsky (Ed.), *Teaching technical communication: Critical issues for the classroom.* Bedford/St.Martin's.

Society for Technical Communication. (n.d.). Defining technical communication. https://www.stc.org/about-stc/defining-technical-communication/

Recommended Resources

Andrews, D., & Roberts, D. (2017, August). Academic makerspaces: Contexts for research on interdisciplinary collaborative communication.
In *Proceedings of the 35th ACM International Conference on the Design of Communication* (pp. 1–7). ACM. https://doi.org/10.1145/3121113.3121230

Johnson-Eilola, J., & Selber, S. A. (2022). Technical communication as
assemblage. *Technical Communication Quarterly*, 1–19. https://doi.org/10.108
0/10572252.2022.2036815

Sherrill, J. T. (2014). *Makers: Technical communication in post-industrial participatory communities.* Doctoral dissertation. Purdue University. https://www.proquest.com/
docview/1666815374?pq-origsite=gscholar&fromopenview=true

Shivers-McNair, A. (2020). Mediation and boundary marking: A case
study of making literacies across a makerspace. *Learning, Culture and Social
Interaction, 24*, 100290. https://doi.org/10.1016/j.lcsi.2019.02.015

Tham, J. (2021). *Design thinking in technical communication: Solving problems through
making and collaboration.* Routledge.

See also: Project Management, Information Design, User Experience

75. Tools

Jacob Craig

Tor makers and designers, tools may most readily call to mind the wide range of technologies and materials available to produce texts and objects. In this sense, the keyword *tools* signify a wide range of objects: 3D printers, scanners, chisels, knitting needles, lathes, document design software, code editing software, and anything else human-designed and human-made used in service to creating other texts and objects.

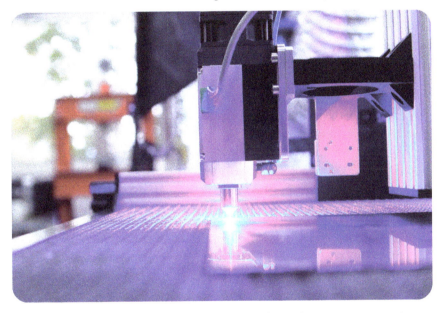

Fig. 75.1. Computer-numerically controlled (CNC) laser cutting tool. Photo by Opt Lasers on Unsplash. https://bit.ly/keyword-fig-75-1

Calling attention to such a wide range of objects as tools has the advantage of suggesting what expertise in design and making entails: knowledge of possible tools available for a particular task; awareness of the features of each available tool; and knowledge of how to use each available tool. An effect of this kind of tool-focused expertise has animated the rise of makerspaces across the country, leading to a resurgence to the democratization of making that parallels the arts and crafts movement of the nineteenth century (Breaux, 2017). Thus, and on one hand, this broadly-focused view of tools proves useful for fundraising and accumulating resources in makerspaces—right tool for the job—or for helping to imagine how particular objects can be used—what a tool affords—in service to a task.

On the other hand, this broadly focused notion of anything human-designed and human-made as a tool also tends to prioritize the means of making/designing over the role of makers and designers. In short, while tools as a keyword denote the technologies of making and design "at the technological, concrete level," it also "suggests a relatively simplistic approach" to understand the means of making and design (Johnson-Eilola & Selber, 2009, p. 21). This simplification happens much in the same way that DIY wrongly suggests that making is not a community-based or public activity but a solitary action (Murphy, Phillips, & Pollock, 2014; Orton-Johnson, 2014). Viewing all technologies as instruments abstracted from their contexts of design, learning, and use occludes particular realities important for effective and responsible making and design (Porter, 2003). Left out of a strictly tool-focused view of hardware, software, machines, and (yes) tools are concerns like how tools can or should or must work together in contexts of making (Porter, 2003); what consumers must have and know to interact with objects and artifacts designed and made with tools (Eyman, 2009); what particular tools were designed to do and for whom (Edwards, 2018); and what a tool can do beyond the intention of the original designer (Leu et al., 2004).

A more contextual view of tools provides an opportunity to understand a particular tool as part of a larger constellation of human and non-human entities that shape each moment of making and designing. For makers and designers, a contextualist view of tools invites expertise in not just how to use particular tools but also how the use of particular tools affects end users, other makers, communities, and even the larger economy. While tool remains an intensely pragmatic way of conceptualizing the materials, machines, and technologies involved in making, it also lends itself to a more social view of making and designing that includes tool makers, tool users, product consumers—as well as other tools (and their makers, users, and consumers).

References

Breaux, C. (2017). Why making? *Computers and Composition, 44*(1), 27–35. https://doi.org/10.1016/j.compcom.2017.03.005

Edwards, D. (2018). Circulation gatekeepers: Unbundling the platform politics of YouTube's content ID, *Computers and Composition, 47*(1), 61–74. https://doi.org/10.1016/j.compcom.2017.12.001

Eyman, D. (2009). Usability: Methodology and design practice for writing processes and pedagogies. In S. Miller-Cochran and R. Rodgrigo (Eds.), *Rhetorically rethinking usability* (pp. 213–228). Hampton Press.

Johnson-Eilola, J., & Selber, S. (2009). The Changing shapes of writing: Rhetoric, new media, and composition. In A.C. Kimme-Hea (Ed.), *Going wireless: A critical exploration of wireless and mobile technologies for composition teachers and researchers* (pp. 15–34). Hampton Press.

Leu, D. J., Kinzer, C. K., Coiro, J. L., & Cammack, D. W. (2004). Toward a theory of new literacies emerging from the internet and other information and communication technologies. In R. B. Ruddell and N. J. Unrau (Eds.), *Theoretical models and processes of reading* (5th ed., pp. 1570–1613). International Reading Association.

Murphy, M., Phillips, D. J., Pollock, K., & Boler, M. (2014). Doing it in the cloud: Google, Apple, and the shaping of DIY culture. In M. Ratto (Ed.), *DIY citizenship: Critical making and social media* (pp. 249–257). MIT Press.

Orton-Johnson, K. (2014). DIY citizenship, Critical making, and community. In M. Ratto and M. Boler (Ed.), *DIY citizenship: Critical making and social media* (pp. 141–155). MIT Press.

Porter, J. (2003). Why technology matters to writing: A cyberwriter's tale, *Computers and Composition, 20*(4), 375–394. https://doi.org/10.1016/j.compcom.2003.08.020

249

Recommended Resources

Einarsson, Á. M., & Hertzum, M. (2021). How do makers obtain information for their makerspace projects? *Journal of the Association for Information Science and Technology, 72*(12), 1528–1544. https://doi.org/10.1002/asi.24528

Gierdowski, D., & Reis, D. (2015). The MobileMaker: An experiment with a mobile makerspace. *Library Hi Tech, 33*(4), 480–496.

Hira, A., & Hynes, M. M. (2018). People, means, and activities: A conceptual framework for realizing the educational potential of makerspaces. *Education Research International*, Article ID 6923617. https://doi.org/10.1155/2018/6923617

Keune, A., & Peppler, K. (2019). Materials to develop with: The making of a makerspace. *British Journal of Educational Technology, 50*(1), 280–293. https://doi.org/10.1111/bjet.12702

Litts, B. K. (2015, June). Resources, facilitation, and partnerships: Three design considerations for youth makerspaces. In *Proceedings of the 14th International Conference on Interaction Design and Children* (pp. 347–350). https://doi.org/10.1145/2771839.2771913

See also: Materiality, Digital Rhetoric, Coding, Repair, Safety

76. Universal Design

Ada Hubrig

The concept of universal design (UD) comes to the maker movement from its origin in architecture and civil engineering. Just as the outline of the person in a wheelchair has become the international symbol for disability and accessibility, the example of the dropped curb (where the sidewalk is graded down to form a ramp to the adjoining street) has become the primary example of UD. The drop curb (also called a curb cut) was the achievement of architect Selwyn Goldsmith. Goldsmith's (1963) architectural book *Designing for the Disabled* was the first text to deal in depth with how to design spaces for people with disabilities. The term "universal design" itself was coined by architect Ronald Mace in 1985 (see Hamraie, 2017).

The ideal of UD stresses that design considerations toward accessibility make the space more open to all. For example, the drop curb not only makes sidewalks more accessible to wheelchair users, but also parents with strollers or people with carts. But the ideal of UD is not without critique: Rick Godden and Jonathan Hsy (2016), in their MLA Position Paper, explore a number of these threads of critique, which is a helpful resource for those considering issues of UD—though the primary critique of UD is that it is simply too idealistic and ultimately unachievable. This critique is concerned with the problematic focus on "everyone" in the language of UD rather than specific needs of specific populations of people.

Throughout *Building Access*, Aimi Hamraie (2017) discusses at length how UD reinforces ableist normativities. Hamraie offers a useful critique of UD, highlighting UD's limitations while centering how people with disabilities are not to be seen as a challenge to overcome in design, but instead people with disabilities are a valuable cultural identity, a source of knowledge, and a basis for relationality (p. 12). In other words, people with disabilities are more than the access/accessibility needs that UD may reduce them to. Many disability studies scholars have critiqued the idealism and limitations of UD while simultaneously acknowledging UD's potential. Jay T. Dolmage (2017) for example, insists that true UD is not an end goal, but a verb, an ongoing process. Rather than focusing on creating the ultimate, accessible-for-everyone product, designers can focus their attention on continually improving the product as the user base grows.

And this version of UD—the ongoing process of improving access—provides a challenging imperative for makers of all stripes. *The Maker Manifesto* argues the importance of ensuring access, stating "One of the keys to [the maker] movement is the democratizing impact of access to the tools one needs to make things" (Hatch, 2013, p. 7). Although not written *about* disability, in thinking about issues of accessibility, the idea nonetheless points to the democratizing impact access can have. Peter Suber (2013) includes handicap access barriers as one of four major barriers open access movements have yet to adequately address. UD—as an ongoing process—imagines access for those sidelined by barriers, democratizing spaces and experiences for access, but UD—and the thoughtful critiques of UD offered to us by people with disabilities—push design to engage disability more deeply, to see disabled embodiment as a source of knowledge-making and world-building.

References

Dolmage, J. (2017). *Academic ableism: Disability and higher education*. University of Michigan Press.

Godden, R., & Hsy, J. (2016). Universal design and its discontents. *2016 MLA Position Papers*. https://www.disruptingdh.com/universal-design-and-its-discontents/

Goldsmith, S. (1963). *Designing for the disabled: A manual of technical information*. Royal Institute of British Architects / Routledge.

Hamraie, A. (2017). *Building access: Universal design and the politics of disability*. University of Minnesota Press.

Hatch, M. (2014). *The maker movement manifesto: rules for innovation in the new world of crafters, hackers, and tinkerers*. McGraw-Hill Education.

Suber, P. (2013). *Open access*. MIT Press. https://mitpress.mit.edu/books/open-access

Recommended Resources

Barbarin, I. (2019, Feb 13). Accessibility isn't charity—it's a lifesaving responsibility. *The Philadelphia Inquirer*. https://www.inquirer.com/opinion/commentary/septa-disability-access-subway-death-20190213.html

Centre for Excellence in Universal Design. (2014). What is universal design? National Disability Authority. http://universaldesign.ie/What-is-Universal-Design/

Centre for Excellence in Universal Design. (2014). Case studies and examples. National Disability Authority. http://universaldesign.ie/What-is-Universal-Design/Case-Studies-and-Examples/

Guffey, E. (2018). *Designing disability: Symbols, space, and society*. Bloomsbury Visual Arts.

Seo, J., & Richard, G. T. (2021). SCAFFOLDing all abilities into makerspaces: A design framework for universal, accessible and intersectionally inclusive making and learning. *Information and Learning Sciences*, *122*(11/12), 795–815. https://doi.org/10.1108/ILS-10-2020-0230

Steele, K., Blaser, B., & Cakmak, M. (2018). Accessible making: Designing makerspaces for accessibility. *International Journal of Designs for Learning*, *9*(1), 114–121. https://www.learntechlib.org/p/209688/

Williamson, B. (2019). *Accessible America: A history of Disability and Design*. New York University Press.

See also: Inclusive Design, Accessibility, Learning Diversity

77. User Experience

Dennis Cheatham

Drinking out of a coffee cup, mapping a destination on a smartphone, or checking in for a flight at an airport ticket counter—all of these activities can affect someone's mood, performance of tasks, and overall satisfaction with the activity. In design, the consideration of an individual's feelings and reception of the experience with an activity, mediated by an interface, is known as user experience. Some user experiences can be undesirable or even harmful. Designers refer to these as negative experiences. For example, an older adult with arthritis could have a painful experience using a spatula because the handle was not designed for people who need a larger grip. Other times, the user experience may be positive, such as when a travel writer thinks a writing app is intuitive to use because the app's features simplify the writing and distribution process while on location. Positive user experience itself can be touted as a reason to select a product. People use products, services, and systems to accomplish tasks, and the quality of the user experience while using a designed outcome defines a successful user experience (Norman & Nielsen, n.d.).

When creating products, services, and systems, designers emphasize user experience at different levels on a broad spectrum. Functionality can take precedence when a product's smooth and accurate operation is of primary importance. Pumping fuel into a car at a gas station emphasizes safety and efficiency, but it's hardly something that makes drivers seek out the best gas pumping experience. On the other end of the spectrum, some design outcomes aspire to create a holistic user experience. These surpass basic functionality and are designed to make people feel things like luxury, delight, or playfulness. A spa intended to create a serene user experience may feature low lighting, extra-soft towels, fluffy robes, and cucumber-infused water. While these details are not necessary for a successful spa treatment, they facilitate a specific kind of user experience.

Mainly, user experience consists of three factors: a user, a designed outcome, and a context where the user interacts with the design (Forlizzi & Ford, 2000). A person's background, physical characteristics, values, preferences, and needs affect how they will respond to designed products. A person who prefers comfort over fashion will select shoes that feel good, sacrificing good looks. The design of a product, service, or system af-

253

fects how useful, usable, and desirable it will be to different users (Sanders, 1992). Complex services that require many steps to complete, like filing taxes, can cause frustration. The places where people use design can drastically affect their experience. Weather conditions, acceptable behaviors, and language differ based on location, such as a synagogue, driving in a car, or a honky-tonk bar in Tulsa. When a product's design is in harmony with the user's makeup and functions within the context's limitations, the user experience will most likely be positive.

References

Forlizzi, J., & Ford, S. (2000). The building blocks of experience: An early framework for interaction designers. *Proceedings from 3rd conference on Designing interactive systems: processes, practices, methods, and techniques* (DIS '00).

Norman, D., & Nielsen, J. (n.d.). The definition of user experience (UX). https://www.nngroup.com/articles/definition-user-experience/

Sanders, E. B.-N. (1992). Converging perspectives: product development research for the 1990s. *Design Management Journal, 3*(4), 49–54.

Recommended Resources

Garrett, J. J. (2011). *The elements of user experience: user-centered design for the Web and beyond.* (2nd ed.). New Riders.

Janlert, L.-E., & Stolterman, E. (2017). *Things that keep us busy: The elements of interaction.* MIT Press.

Norman, D. A. (2013). *The design of everyday things* (Revised and expanded edition ed.). Basic Books.

O'Donovan, C., & Smith, A. (2020). Technology and human capabilities in UK makerspaces. *Journal of Human Development and Capabilities, 21*(1), 63–83. https://doi.org/10.1080/19452829.2019.1704706

Schifferstein, H. N. J., & Hekkert, P. (Eds.). (2008). *Product experience* (1st ed.). Elsevier.

See also: User Requirements, User Stories, Technical Communication

78. User Requirements

Joseph Bartolotta

Before starting any user design task, it is important for designers to identify what sorts of tasks users are trying to accomplish when they are using an interactive system. This requires designers to assess user needs and then develop requirements that the interactive system must be able to satisfy before its release. User requirements are identified as part of the human-system interaction as described in the international standards, ISO 9241–210: Ergonomics of human-system interaction–Part 210: Human-centered design for interactive systems (https://www.iso.org/standard/77520.html).

User needs tend to not mention the interactive system. This helps designers avoid seeing the system itself as a solution, and instead allows the user requirement to offer a heuristic for determining whether or not the system succeeds or fails in its objective.

Here is a sample user need: College students looking up classes they plan to register for need to know what options they have with regard to the time and where the course is being held so it does not interfere with other courses they have registered for. The user requirement could then be:

1. Users must be able to identify the time and location of the courses they are looking at.
2. Users must be able to tell if a course they are looking at conflicts with a course they have already registered for.

User requirements, while helpful, are not the only requirements designers need to consider. There may also be marketing requirements (i.e., dictates on using certain branding, typefaces, or colors), or organizational requirements (i.e., users may not sign up for a particular course if they have not completed a prerequisite). These latter two requirements are seldom evaluated in usability testing, unless adhering to them results in a massive system failure, at which point they and the design must be revisited and revised. User requirements may be qualitative or quantitative in scope. Using the requirements described above, designers may devise user requirements that may be tested later in usability testing.

Qualitative: Users must be able to identify the time and location of the courses they are looking at; Users must be able to tell if a course they are looking at conflicts with a course they have already registered for.

Quantitative: Ninety-five percent of users must be able to identify the time and location of a course in fifteen seconds. Eighty-five percent of users must answer that they "agree" or "strongly agree" to the statement: "I would recommend this interface to a friend."

These sorts of requirements are helpful both in the design stage and offer some specific points of investigation during usability testing. They allow designers to break down components of their work into more manageable objectives.

Recommended Resources

Cooper, A., Reimann, R., Cronin, D., & Noessel, C. (2014). *About face: The essentials of interaction design (4th edition)*. John Wiley & Sons, Inc.

Courage, C., & Baxter, K. (2015). *Understanding your users: A practical guide to user requirements methods, tools, and techniques (2nd edition)*. Morgan Kaufmann.

Maguire, M., & Bevan, N. (2002, August). User requirements analysis. In *IFIP World Computer Congress, TC 13* (pp. 133–148). Springer.

Mayhew, D. J., & Mayhew, D. (1999). *The usability engineering lifecycle: a practitioner's handbook for user interface design*. Morgan Kaufmann.

Wood, L. E. (1997). *User interface design: Bridging the gap from user requirements to design*. CRC Press.

See also: User Experience, User Stories, Context & Contextual Design

79. Writing Studio

Matthew Kim & JongHun Kim

For some compositionists, the makerspace is a studio version of writing pedagogy. The concept of writing studio has evolved over the past two decades as a space/place where participants are afforded unique opportunities to explore their writing, designing, and making outside but alongside the classroom. At the heart of the writing studio is the mantra "learn through doing." The writing studio model was piloted by Nancy Thompson and Rhonda Grego in the 1990s at the University of South Carolina, as a space/place where students labeled basic writers could meet as a group with a facilitator to discuss academic writing and the writing process. The writing studio is similar to the fine arts studio in that "apprentice learners master their craft through directed group participation" (Sutton & Chandler, 2018).

Thompson and Grego have established both a physical writing studio and a writing studio methodology of interactional inquiry. Interactional inquiry "functions as action research" that is "made up of conversation and writing produced through an on-going series of reflective interactions where participants examine and rethink their experiences" (Sutton & Chandler, 2018). This writing studio methodology affords participants and facilitators opportunities to critique dominant norms about the ways in which writing is—or is not—valued at an institution and looks for ways to help "unheard voices speak into the silences more immediately and directly located within their own institutional spaces/places" (Grego & Thompson, 2008 p. 188). Engaging in interactional inquiry in a writing studio where participants are designing multimodal texts and fabricating 3D objects positions participants to examine specific tools of production and distribution; how our environment impacts the production and distribution of texts and objects; and how people interact as they produce, distribute, and use the texts and objects that participants make.

Other writing studio researchers and practitioners important to the development of the writing studio model across institutional contexts include S. Morgan Gresham (2010), James A. Inman (2013), and Ben Lauren (2013). Russell Carpenter (2013, 2014, 2016), director of the Noel Studio at Eastern Kentucky University (https://studio-old.eku.edu/), adds to the writing studio narrative, claiming that the significance of a physical space matters as much as the theoretical space in which the studio emerges. Carpenter argues that writing studio pedagogy drives space design; that studios focus on spatial designs and the flow of student traffic; and that an interactive approach to consulting with students on the design of highly effective communication products, considering students' multiliteracy skills and learning styles through integrated practices. Matthew Kim (2019) considers how the rituals with which students and facilitators engage in the writing studio, including conversation, divergent thinking, spatial awareness, and play, lead them towards becoming rhetorically savvy writer-designer-performers.

For example, in Kim's studio participants practice these rituals to invent potential arguments for their writing and making projects using Augusto Boal's *Theatre of the Oppressed* categories of games and exercises, as well as his more well-known Forum Theatre. Participants, tapping into their imagination, use their bodies to perform roles of oppressors and liberators in different situations. Kim fashions Boal's games for studio participants in ways that engage them in thinking about composing in particular genres, such as photo essays and researched arguments. These performances move participants in the writing studio towards discovering what might want to be expressed in their compositions, providing them opportunities to think about writing and making beyond classical or social-epistemic invention.

As more students are required to produce multimodal compositions and 3D objects in their composition courses and do their making on cloud-based tools, writing studios must shift from a merely a collaborative culture to a do-it-with-others (DIWO) culture, or one that engages those in the studio in discussions and activities on their writing, designing, and making as co-produced, networked activities. JongHun Kim (2017) implements a writing studio model in his middle and high school music courses in which he emphasizes DIWO culture and informal learning. In his music studio, participants bring their writing and making projects from other classes and explore and reflect upon those projects through musical activities. He asserts that creativity blooms with informal learning in the writing studio as participants voluntarily engage in harmonizing, finding a rhythm, or coming together as an ensemble to solve a problem.

When writing studio participants are afforded opportunities to design their own spaces and have creative control over their invention and problem-solving activities, their interest in writing and making develop positively. As well, participants involved in a writing studio learn knowledge and skills that have the potential to transfer with them across their academic lives. It is important to remember that the writing studio exists outside and alongside the makerspace and other more traditional learning environments. Therefore, writing studio directors and facilitators, makerspace directors, and writing program administrators must communicate openly and inquire frequently about each other's needs to ensure student and department successes.

References

Carpenter, R., Valley, L., Napier, T., & Apostel, S. (2013). Studio pedagogy: A model for collaboration, innovation, and space design. In R. Carpenter (Ed.), *Cases on higher education spaces; Innovation, collaboration, and technology* (pp. 313–329). IGI Global.

Carpenter, R. (2014). Negotiating spaces of design in multimodal composition. *Computers and Composition: An International Journal, 33*(1), 68–78.

Carpenter, R. (2016). Shaping the future: Writing centers as creative multimodal spaces. *SDC: A Journal of Multiliteracy and Innovation, 21*(1), 56–75.

Chandler, S., & Sutton, M. (2018). Writing studios and change. In M. Sutton and S. Chandler (Eds.,), *The writing studio sampler: Stories about change* (pp. 3–26). The WAC Clearinghouse and University Press of Colorado.

Grego, R., & Thompson, N. (2008). *Teaching/Writing in thirdspaces: The studio model.* Southern Illinois University Press.

Gresham, M. (2010). Composing multiple spaces: Clemson's class of '41 online studio. In D.M. Sheridan and J.A. Inman (Eds.), *Multiliteracy centers: Writing center work, new media, and multimodal rhetoric* (pp. 33–55). Hampton Press.

Inman, J. (2010). Designing multiliteracy centers: A zoning approach. In D.M. Sheridan and J.A. Inman (Eds.), *Multiliteracy centers: Writing center work, new media, and multimodal rhetoric* (pp. 19–32). Hampton Press.

Kim, J. (2017). Enacting writing studio pedagogy in the music studio." In M. Kim and R. Carpenter (Eds.), *Writing studio pedagogy: Space, place, and rhetoric in collaborative environments* (pp. 81–92). Rowman & Littlefield.

Kim, M. (2019). Performing the writing studio in the inclusive classroom: Asserting a multimodal language with which to contest the learning disabilities label at school. In R. Carpenter (Ed.), *Studio-based approaches for multimodal projects: Models to promote engaged student learning* (pp. 53–74). Rowman & Littlefield.

Lauren, B. (2013). Designing small spaces: A case study of the Florida International University Digital Writing Studio. In R. Carpenter (Ed.), *Cases on higher education spaces; Innovation, collaboration, and technology* (pp. 64–86). IGI Global.

Recommended Resources

Brooks-Gillies, M., Garcia, E. G., & Manthey, K. G. (2020). Making do by making space: Multidisciplinary graduate writing Groups as spaces alongside programmatic and institutional places. In M. Brooks-Gillies, E. G. Garcia, S. H. Kim, K. Manthey, & T. G. Smith (Eds.), *Graduate writing across the disciplines: Identifying, teaching, and supporting* (pp. 191–209). The WAC Clearinghouse/University Press of Colorado. https://doi.org/10.37514/ATD-B.2020.0407.2.08

Gough, N. (1992). Laboratories in schools: Material places, mythic spaces. Paper presented at the Annual Meeting of the American Educational Research Association (San Francisco, CA, April 20–24, 1992). https://files.eric.ed.gov/fulltext/ED356955.pdf

Lammers, J. C., Magnifico, A. M., & Curwood, J. S. (2014). Exploring tools, places, and ways of being: Audience matters for developing writers. In K.E. Pytash and R.E. Ferdig (Eds.), *Exploring technology for writing and writing instruction* (pp. 186–201). IGI Global.

Reynolds, N. (2007). *Geographies of writing: Inhabiting places and encountering difference.* Southern Illinois University Press.

Sabatino, L. (2017). Fostering writing studio pedagogy in space designed for digital composing practices. In M. Kim and R. Carpenter (Eds.), *Writing studio pedagogy: Space, place, and rhetoric in collaborative environments* (pp. 177–197). Rowman & Littlefield.

See also: Composition Commons, Inter/Cross-Disciplinarity, Mentoring, DIWO

PART 3:
METHODS

80. 3D Printing

Charles Woods

The term "3D printing" refers to using a specialized printer to produce a three-dimensional (3D) object. 3D printing is a popular digital technology occupying makerspaces and represents the relationship between digital technology and material production within maker culture. The initial purpose of the technology upon its inception in the 1980s was *rapid prototyping*; that is, to produce prototypes efficiently (quickly and cheaply). Since Carl Deckard developed layered manufacturing and printed a 3D model in 1987 by utilizing laser light for fusing the metal powder in solid prototypes, rapid prototyping has covered vast territories from the medical field, to the aviation sector, to artful sculpting (Bagaria, Rasalkar, Bagaria, & Ilyas, 2011). 3D printing is referred to as *additive manufacturing* due to the way in which the technology functions. In opposition to subtractive manufacturing, additive manufacturing like 3D printing continuously adds material layers until the model being printed is produced.

Fig. 80.1. A RepRap 3D printer in action. Photo by Gavin Allanwood on Unsplash. https://bit.ly/keyword-fig-80–1

Although there are variations to 3D printing technologies, it is helpful to understand how this unique technology works. Ultimately, thin layers of plastic filaments are laid on top of the next until the object is produced. The plastic filaments are fed into the printer and melted in the print head, which excretes the filament onto the build plate to produce the object. Producing a 3D printed object is not immediate; the time it takes a 3D printer to produce a 3D object from a digital model depends upon a variety of factors, including the quality of the printer, the intricacies of the design, and the mass of the object to be produced.

The literacies needed to produce a 3D-printed object do not end with the printer technology; makers must also be familiar with computer-aided design (CAD) programs. Instructors who wish to implement 3D printing in the classroom must make sure to account for instruction concerning CAD programs, but many online programs do offer support in the form of tutorials and resources. Makers can use CAD programs, like TinkerCad (Tinkercad.com) or FreeCAD (freecadweb.com), to construct a digital model of the object to be printed. Beginner makers might find TinkerCad's interface and resources helpful; seasoned makers might find FreeCAD beneficial to their needs, as FreeCAD offers an archive of models to be adjusted and amended for printing.

Access is of central concern to any conversation concerning technology, including conversations within and about maker culture. Using free design templates and designs available under Creative Commons shows one small way that the movements and communities surrounding 3D printing can be a more inclusive space (Wysocki, 2017). At a time when there always seems to be some new way to compose, it is good to know that these technologies are understandable, can be learned, and can be employed (Alvarez et al., 2017). Ultimately, 3D printing, and other makerspace technologies, helps students on their way toward the greater responsibilities of educated citizens (Kill, 2012) by providing opportunities that will help them to develop both understanding of the collaborative nature of knowledge-making and skill in learning and working collaboratively.

References

Alvarez, S. P., Baumann, M., Day, M., Echols, K. L., Gordon, L. M. P., Kumari, A., Matravers, L. S., Newman, J., Nichols, A. M., Ray, C. E., Udelson, J., Wysocki, R., & DeVoss, D. N. (2017). On multimodal composing. *Kairos: A Journal of Rhetoric, Technology, and Pedagogy, 21*(2). http://kairos.technorhetoric.net/21.2/praxis/devoss-et-al/index.html

Bagaria, V., Rasalkar D., Bagaria S. J., & Ilyas, J. (2011). Medical applications of rapid prototyping. In M. Hoque (Ed.), *A new horizon:*

Advanced applications of rapid prototyping technology in modern engineering (pp. 1–20). InTech. http://www.intechopen.com/books/advanced-appli-cations-of-rapid-prototyping-technology-in-modern-engineering/medical-applications-of-rapid-prototyping-a-new-horizon

Kill, M. (2012). Teaching digital rhetoric: Wikipedia, collaboration, and the politics of free knowledge. In B. D. Hirsch (Ed.), *Digital humanities pedagogy: Practices, principles, and politics* (pp. 389–405). Open Book Publishers. http://books.openedition.org/obp/1658

Recommended Resources

Chee, K. C., & Leong, K. F. (2017). *3D printing and additive manufacturing: Principles and applications* (5th edition). World Scientific Publishing Co.

Choong, Y. Y. C., Tan, H. W., Patel, D. C., Choong, W. T. N., Chen, C.-H., Low, H. Y., Tan, M. J., Patel, C. D., & Chua, C. K. (2020). The global rise of 3D printing during the COVID-19 pandemic. *Nature Reviews Materials, 5*(9), 637–639. https://doi.org/10.1038/s41578-020-00234-3

Ford, S., & Minshall, T. (2019). Invited review article: Where and how 3D printing is used in teaching and education. *Additive Manufacturing, 25*, 131–150. https://doi.org/10.1016/j.addma.2018.10.028

Lipson, H., & Kurman, M. (2013). *Fabricated: The new world of 3D printing*. John Wiley & Sons.

Shahrubudin, N., Lee, T. C., & Ramlan, R. J. P. M. (2019). An overview on 3D printing technology: Technological, materials, and applications. *Procedia Manufacturing, 35*, 1286–1296.

See also: Physical Computing, Makerspace, Maker Faire, Hacking

81. A/B Testing

Halcyon M. Lawrence

In interactive design, A/B testing is a usability research method that uses online controlled experiments. In the experiment, some change in a variable (treatment) is presented randomly to users to determine if that change in variable performs better than the unchanged (control) variable. For example, a research team for an online store might test to see if US-based customers are more likely to click if a "Buy Now" button is red (treatment) or green (control). A/B testing often serves a conversion goal, which might be measured by a purchase, longer user interaction with the site, or a user becoming a new customer, for example.

Research employing A/B testing is used in a range of domains including games (Francisco-Aparicio et al., 2013), recommender systems (Gilotte et al., 2018), online stores (Siroker & Koomen, 2013), search engines (Kohavi & Longbotham 2016), and Massive Open Online Courses (Renz, 2016).

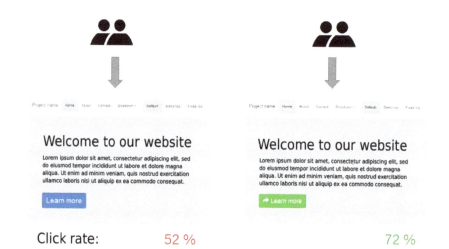

Fig. 81.1. An example of A/B testing for click rate between two designs. Figure by Maxime Lorant via Wikimedia Commons. https://commons. wikimedia.org/wiki/File:A-B_testing_example.png

In A/B testing of websites, tests are done live with visitors to the site. Visitors are randomly and evenly placed between the treatment and control groups. Researchers control for other external variables, like time of day, seasonality, or location, so that any difference in performance between the two groups is attributed to the change in the variable or random chance; the latter is accounted for through significance testing (Fabijan 2018). Given the growing success of A/B testing (Siroker & Koomen, 2013), a number of best practices have emerged, which include:

- Having a clear idea of what is being optimized (i.e., quantifiable success metrics);
- Presenting true randomized condition to users;
- Testing a single change in a variable at a time;
- Using a sufficient population size;
- Choosing an appropriate statistical significance level; and
- Interpreting correctly the statistical significance of findings.

Despite its effectiveness, A/B testing has a few drawbacks. First, while the results indicate user preferences, researchers won't understand the motivations that lead users to these preferences. Follow-up interviews and focus groups might be some ways to mitigate this limitation. Second, the high costs associated with running A/B tests may be a barrier to use, especially for smaller companies and non-for-profit organizations. Recently, there have been services like Optimizely (https://www.optimizely.com/) that offer online tools for A/B testing to organizations that do not have in-house website research teams. Finally, A/B testing is an iterative research process that requires a significant investment of time by research teams (Gilotte et al., 2018).

There are a few reasons that A/B testing as a research method aligns with the goals and practices of design thinking. First and perhaps most important, given design thinking's human-centered focus (Owen, 2017), A/B testing allows researchers to collect actual user data, helping to move design decisions away from intuition and guess work (Fabijan 2018). Second, given design thinking's affinity for teamwork, A/B testing is by nature collaborative and interdisciplinary (Siroker & Koomen, 2013), pulling together teams of coders, designers, researchers, and marketers, and of course, users. Therefore, research teams interested in employing methodologies aligned with design thinking might find A/B testing a useful research method.

References

Fabijan, A., Dmitriev, P., Olsson, H. H., & Bosch, J. (2018, August). Online controlled experimentation at scale: An empirical survey on the current state of A/B testing. In *2018 44th Euromicro Conference on Software Engineering and Advanced Applications (SEAA)* (pp. 68–72). IEEE.

Francisco-Aparicio, A., Gutiérrez-Vela, F. L., Isla-Montes, J. L., & Sanchez, J. L. G. (2013). Gamification: Analysis and application. In V. M. R. Penichet, A. Peñalver, and J. A. Gallud (Eds.), *New trends in interaction, virtual reality and modeling* (pp. 113–126). London, UK: Springer.

Gilotte, A., Calauzènes, C., Nedelec, T., Abraham, A., & Dollé, S. (2018, February). Offline A/B testing for recommender systems. In *Proceedings of the Eleventh ACM International Conference on Web Search and Data Mining* (pp. 198–206). ACM.

Kohavi, R., & Longbotham, R. (2017). Online controlled experiments and A/B testing. In C. Sammut and G. I. Webb (Eds.), *Encyclopedia of machine learning and data mining*, (pp. 922–929) Springer.

Owen, C. (2007). Design thinking: Notes on its nature and use. *Design Research Quarterly*, 2(1), 16–27.

Renz, J., Hoffmann, D., Staubitz, T., & Meinel, C. (2016, April). Using A/B testing in MOOC environments. In *Proceedings of the Sixth International Conference on Learning Analytics & Knowledge* (pp. 304–313). ACM.

Siroker, D., & Koomen, P. (2013). *A/B testing: The most powerful way to turn clicks into customers.* John Wiley & Sons.

Recommended Resources

Georgiev, G. (n.d.). Statistical significance in A/B testing—a complete guide. Analytics toolkit. http://blog.analytics-toolkit.com/2017/statistical-significance-ab-testing-complete-guide

Gilotte, A., Calauzènes, C., Nedelec, T., Abraham, A., & Dollé, S. (2018). Offline A/B testing for recommender systems. In *Proceedings of the Eleventh ACM International Conference on Web Search and Data Mining* (pp. 198–206). https://doi.org/10.1145/3159652.3159687

Gui, H., Xu, Y., Bhasin, A., & Han, J. (2015, May). Network A/B testing: From sampling to estimation. In *Proceedings of the 24th International Conference on World Wide Web* (pp. 399–409). https://doi.org/10.1145/2736277.2741081

Khandelwal, A. (2023). 7 A/B testing examples to bookmark. VWO. https://vwo.com/blog/10-kickass-ab-testing-case-studies/

Kohavi, R., Tang, D., & Xu, Y. (2020). *Trustworthy online controlled experiments: A practical guide to A/B testing.* Cambridge University Press.

See also: User Experience, User Requirements, Coding

82. Affinity Diagramming

Arthur Berger

To understand how users think or feel about certain designs, makers collect information regarding user requirements or feedback as data that can inform their next decisions. To organize the collected data, designers need a systematic method. Affinity diagramming is a way to organize things into groups, based upon their natural relationships. To make an affinity diagram, list out items and then sort the items into groups, where each item of the group is similar to other items in the group. For example, you might write each item on an individual sticky note and then rearrange the sticky notes into different groups on a whiteboard. If you have a data set with hundreds of items, you might combine groups. When every item belongs to a group, you are done sorting and can label the groups with a descriptive name. Now, you have an affinity diagram of the categories or themes (the groups) of your data set (the list of items) (Tague, 2005).

In the mid twentieth century, Kawakita Jiro, a Japanese anthropologist of East Asian ethnogeography studies, classified a variety of human and natural information in order to clearly communicate the unique eco-cultural practices of a certain area to people generally unfamiliar with the subject matter. Jiro's "let the chaos speak for itself" research method, with its focus on practical problem-solving from a large set of different data types, became known as the KJ Method, and was adopted by business engineering and quality control process improvement disciplines such as Lean Six Sigma ("Kawakita Jiro," n.d.). As manufacturing companies around the world adopted the Japanese "total quality control" management practices at the turn of the century, the KJ Method began to be used more often in other disciplines such as human factors and product design, where it became known more descriptively as affinity diagramming or mapping (Tague, 2005).

As the history of the method suggests, affinity diagramming is used when you have large, complex, or otherwise "chaotic" sets of data or bits of information that you must organize into more manageable groups. Common use cases include verbal data such as from text-based forums, web-scraping, surveys, talk-aloud usability testing protocols, taxonomies and other naming conventions, or a collection of observed behaviors or experiences.

In making and design thinking, the design process moves through several stages, such as empathize > define > ideate > prototype > test (Waloszek, 2012). Each stage itself might have multiple phases, such as *ideation*, which moves through an iterative process of diverging and converging ideas. These stages help address common issues in product design such as lack of critical information about users, user diversity, rigid and ratio-centric planning specifications, and late iteration (Gould & Lewis, 1985). If you have a large amount of research data, such as from user surveys, that you want to guide the empathize or define stages, you can sort the data into more manageable groups with affinity diagramming. Otherwise, this method is most commonly used during the ideate stage, particularly as a way to collect a large set of observations or brainstorm ideas (diverge) and then organize these together (converge). The affinity diagram becomes an input to other design products for use in the prototype and test stages. Although terms like ideate and brainstorming might suggest that affinity diagrams are used for new data and new insights, the method can also be effectively applied to existing information, such as when redesigning an interface.

Fig. 82.1. Color-sorted cards (sticky notes) arranged to show affinity/ categories and meanings. Photo by Hugo Rocha on Unsplash. https://bit. ly/keyword-fig-82–1

By identifying the themes of your data, you create a set of criteria that you can further refine into relational, cause-effect, and network analyses (Lepley, 1998). Some common next steps to take include using the affinity diagram to create as-is or to-be scenarios, user empathy maps, need statements, and prioritization grids for decision-making ("Enterprise design thinking," n.d.). Affinity diagrams are a key component in making sure that these analyses are grounded in the actual, expressed situation that your product, process, or solution is supposed to address, as collected in your initial data set.

While affinity diagramming can be done individually, it is often done in groups. The group might have a moderator or a rule to guide the process, such as everyone brainstorming and arranging ideas silently or separately, allowing participants to rearrange already-positioned ideas, and then discussing as a group to finalize the affinity map (Tague, 2005). Moderation rules can be adjusted based on the team and desired outcome. For example, in a group with hierarchies of position, rules of silence might encourage participation from everyone within the group as a way to discourage bias or top-down directives. Another example might be a placeholder category for contested items and facilitated discussion so that the group negotiates and agrees upon the final outcome (Christensen, 2017).

Common tools that are used to make affinity diagrams include physical tools like pens, notecards, sticky notes, and whiteboards, as well as virtual collaboration tools like LucidChart or Mural. This method is used in diverse fields such as business, engineering, quality control, project management, product and software design, technical communication, health sciences, psychology, human factors, mathematics, artificial intelligence, and education. Related terms include affinity mapping, Kawakita Jiro or KJ Method, and thematic analysis.

References

Christensen, E. (2017). Affinity diagrams: Your key to more creative problem solving. LucidChart. https://www.lucidchart.com/blog/affinity-diagrams-your-key-to-more-creative-problem-solving

"Enterprise design thinking." (n.d.). https://www.ibm.com/design/thinking/

Gould, J. D., & Lewis, C. (1985). Designing for usability: Key principles and what designers think. *Communications of the ACM, 28*(3), 300–311.

"Kawakita Jiro." (n.d.). http://fukuoka-prize.org/en/laureate/prize/acd/kawakita.php

Lepley, J. (1998). Problem-solving tools for analyzing system problems: The affinity map and the relationship diagram. *JONA: The Journal of Nursing Administration, 28*(12), 44–50.

Tague, N. (2005). *Quality toolbox* (2nd edition). American Society for Quality, Quality Press.

Waloszek, G. (2012). Introduction to design thinking. SAP User Experience Community. https://experience.sap.com/skillup/introduction-to-design-thinking/

Recommended Resources

Awasthi, A., & Chauhan, S. (2012). A hybrid approach integrating affinity diagram, AHP and fuzzy TOPSIS for sustainable city logistics planning. *Applied Mathematical Modelling, 36*, 573–584.

Bul, K., Krishnan, N., Griggs, D., Mccarthy, K., Whiteman, B., Szczepura, A., & Numora, T. (2018). Using the Japanese KJ Ho method as a qualitative creative problem solving technique to address clinicians' and young kidney transplant patients' needs concerning treatment. *Transplantation, 102* Suppl 7S-1, S594–S594.

Holtzblatt, K., & Beyer, H. (2017). The affinity diagram. In *Contextual design* (2nd edition) (pp. 127–146). Morgan Kaufmann.

Remy, C., Harboe, G., Frich, J., Biskjaer, M. M., & Dalsgaard, P. (2021). Challenges and opportunities in the design of digital distributed affinity diagramming tools. In *Proceedings of the 32nd European Conference on Cognitive Ergonomics* (pp. 1–5). https://doi.org/10.1145/3452853.3452871

Widjaja, W., Yoshii, K., Haga, K., & Takahashi, M. (2013). Discusys: Multiple user real-time digital sticky-note affinity-diagram brainstorming system. *Procedia Computer Science, 22*, 113–122.

See also: Card Sorting, Case Study, User Requirements, Iterative Design

83. Autoethnography

Erin Kathleen Bahl

Autoethnography is a research methodology in which the researcher is involved in the study as a research subject. This self-generated data may be the main focus of the study or interwoven with others' accounts, but for research purposes is typically framed as one deep dive into significant experiences that speak to cultural phenomena on a broader level.

Autoethnography as methodology has been taken up by multiple disciplines and may overlap in part or in whole with other terms such as creative ethnography, ethnographic memoir, or critical autobiography (Boyle & Parry, 2007; Ellis, 2004; Ellis & Bochner, 2000; Ellis & Bochner, 2016; Franks, 2016; Lawless, 2005). Due to the reflective nature of these studies, autoethnographic research publications sometimes bend the genre of "scholarly essay" in creative, personal, and self-aware ways. These may range from relatively traditional scholarly articles to personal narratives, and from live performances to multimedia pieces (Ellis, 2004; Ronai, 1992; Uotinen, 2010).

Autoethnography is a particularly useful methodology for conducting research on making and design practices when the researchers themselves are also engaged in those practices. These individuals have access to a rich body of self-generated data already available, at a level of depth and intimacy that would be intrusive to solicit from other human subjects. For those engaged in digital making and design, autoethnography is a potentially accessible method of data collection through modes such as screen recordings, video recordings, drafts, and reflective process notes. When employed as a research methodology for making and design, autoethnography can offer significant insight, understanding, and systematicity in framing one's own work, as well as a framework for situating one's own making practices within a network of others pursuing similar work.

One challenge when conducting an autoethnographic study of one's own making practice is to remain highly reflexive in performing design and systematic in studying design processes, while recognizing the ways in which self-observation may shape the data collected. Challenges may also arise in determining data collection parameters and determining how to analyze the potentially enormous amount of information gathered. As an autoethnographer, the researcher is embedded in a massive web of data and may have difficulty stepping back to make sense of it, or finding critical distance from personal experiences.

Several scholar-designers practicing autoethnographic methods in digital making include Alexandra Hidalgo (2016), bonnie lenore kyburz (2019), Jonathan Rhodes and Jacqueline Alexander (2015), and Jody Shipka (2016). These designers' works involve the composers themselves as key presences for reflective data generation, through personal narrative, reflective scholarship connected to broader trends, and even embodied presence through media artifacts. In particular, these scholars draw on the affordances of filmmaking and video production to embed their faces, voices, actions, and surroundings as central features of their overall scholarly arguments. Such research embraces personal experiences as significant dimensions of knowledge creation, and invites audiences to engage versions of those experiences as foundational data sources.

References

Boyle, M., & Parry, K. (2007). Telling the whole story: The case for organizational autoethnography. *Culture and Organization, 13*(3), 185–190.

Ellis, C. (2004). *The ethnographic I: A methodological novel about autoethnography*. Rowman and Littlefield Publishers, Inc.

Ellis, C., & Bochner, A. (2000). Autoethnography, personal narrative, reflexivity: Researcher as subject. In N. K. Denzin and Y. S. Lincoln (Eds.), *Handbook of qualitative research* (pp. 733–768). SAGE Publications.

Ellis, C., & Bochner, A. (2016). *Evocative autoethnography: Writing lives and telling stories*. Routledge.

Franks, T. M. (2016). Purpose, practice, and (discovery) process: When self-reflection is the method. *Qualitative Inquiry, 22*(1), 47–50.

Hidalgo, A. (2016). Family archives and the rhetoric of loss. In P. W. Berry, G. E. Hawisher, and C. L. Selfe (Eds.), *Provocations: Reconstructing the archive*, featuring the work of Erin R. Anderson, Trisha N. Campbell, Alexandra Hidalgo, and Jody Shipka. Utah State University Press/Computers and Composition Digital Press. http://ccdigitalpress.org/reconstructingthearchive/hidalgo.html

kyburz, b. l. (2019). *Cruel auteurism: Affective digital mediations toward film-composition*. The WAC Clearinghouse and University of Colorado Press. https://wac.colostate.edu/books/writing/cruel/

Lawless, E. (2005). Special issue on creative ethnography. *Journal of American Folklore, 118*(467).

Rhodes, J., & Alexander, J. (2015). *Techne: Queer meditations on writing the self*. Computers and Composition Digital Press/Utah State University Press. http://ccdigitalpress.org/techne/

Ronai, C. R. (1992). The reflexive self through narrative: A night in the life of an exotic dancer/researcher. In C. Ellis and M. Flaherty (Eds.), *Investigating subjectivity: Research on lived experience* (pp. 102–124). SAGE Publications.

Shipka, J. (2016). The things they left behind: The estate sale as archive. In P. W. Berry, G. E. Hawisher, and C. L. Selfe (Eds.), *Provocations: Reconstructing the archive* (n.p.). Utah State University Press/Computers and Composition Digital Press. http://ccdigitalpress.org/reconstructingthearchive/shipka.html

Uotinen, J. (2010). Digital television and the machine that goes 'PING!': Autoethnography as a method for cultural studies of technology. *Journal for Cultural Research*, 14(2), 161–175.

Recommended Resources

Bowers, S., Chen, Y.-L., Clifton, Y., Gamez, M., Giffin, H. H., Johnson, M. S., Lohman, L., & Pastryk, L. (2022). Reflective design in action: A collaborative autoethnography of faculty learning design. *TechTrends*, 66(1), 17–28. https://doi.org/10.1007/s11528-021-00679-5

Chang, H., Ngunjiri, F., & Hernandez, K. A. C. (2016). *Collaborative autoethnography*. Routledge.

Cunningham, S. J., & Jones, M. (2005). Autoethnography: A tool for practice and education. In *Proceedings of the 6th ACM SIGCHI New Zealand Chapter's International Conference on Computer–Human Interaction: Making CHI Natural* (pp. 1–8). https://doi.org/10.1145/1073943.1073944

Jackson, R., & McKinney, J. G. (Eds.). (2021). *Self+culture+writing: Autoethnography for/as writing studies*. Utah State University Press.

Jones, S. H., Adams, T. E., & Ellis, C. (2016). *Handbook of autoethnography*. Routledge.

Zhou, D., Gomez, R., Davis, J., & Rittenbruch, M. (2022). Engaging solution-based design process for integrated STEM program development: An exploratory study through autoethnographic design practice. *International Journal of Technology and Design Education*, 1–32. https://link.springer.com/article/10.1007/s10798-022-09745-2

See also: Reflection, Cognitive Dissonance, Fieldwork

84. Bodystorming

Chloe Anna Milligan

Brainstorming is often a generative stage of the writing process, but when design thinking sparks making, just brainstorming may not be enough. As brainstorming can lead to thinking through concepts in the abstract, a more concrete and embodied process strategy is better suited to making tools and texts increasingly interactive in nature. Bodystorming is a technique used in interaction and user experience design that turns thinkers into doers, having them work through the implementation of a product through action and performance. Specifically, it calls for designers to imagine as if their product in mind already exists and act accordingly, most preferably in the space it would be implemented according to the scenario that would call for it. Bodystorming is a form of brainstorming "in the wild" (Oulasvirta, Kurvinen, & Kankainen, 2003).

The term bodystorming was coined by Colin Burns in his work with Eric Dishman, William Verplank, and Bud Lassiter (1994) on what they first called "informance" design. Bringing the performance to information design, informance design, or bodystorming, is meant to be an action-oriented, not just analytical, process that thinks through doing. Written observation in service of brainstorming requires too much complex explanation to fully depict product vision as computers become more sophisticated, and they often abstract the designer away from the site in which users will physically interface with their computational product (Oulasvirta et al., 2003). Brian K. Smith (2014) breaks down the major aspects of bodystorming that combat these problems into three loose categories: design in place, prototype in place, and embodied performance. In other words, bodystorming works best in the place of its intended use, with concrete representations of its product vision, through bodily involvement and commitment to scenarios.

In the context of making and design thinking, it may come as no surprise that bodystorming has long had a computational, even ubiquitous computing, history. The operative example provided by Burns et al. (1994) in their proposal of the term concerned the construction of a computer workstation to convince a hairdresser that did not believe a computer would help her run her business. But perhaps the most famous example of bodystorming in the computer workplace is Jeff Hawkins' inspiration for his invention of the Palm Pilot handheld computer device: a block of wood he carried around everywhere, cut out and shaped to resemble the look and feel of what it would be like to handle his envisioned product (Oppegaard & Still, 2013). In the current smartphone application economy, this technique can continue to come in handy with the wiser design of apps and new products.

Bodystorming should assist designers to create with empathy for the user of their product in mind from the start. In contrast to design approaches that spotlight the technology itself, bodystorming can advance a more technology-supported design approach that centers how the user's body will interact with said technology (Márquez Segura, Turmo Vidal, & Rostami, 2016). Dennis Schleicher, Peter H. Jones, and Oksana Kachur Niedzielski (2010) call for a more nuanced concept beyond bodystorming as they see it codified in the literature, which they slightly differentiate as embodied storming to pay closer attention to the variable experience of physical performance. But, whether one calls it informance design or embodied storming, bodystorming as a keyword for making and design thinking makes interaction and user experience design a full-body commitment.

References

Burns, C., Dishman, E., Verplank, W., & Lassiter, B. (1994). Actors, hairdos & videotape–informance design: Using performance techniques in multi-disciplinary, observation based design. *CHI '94: Conference Companion on Human Factors in Computing Systems* (pp. 119–120).

Márquez Segura, E., Turmo Vidal, L., & Rostami, A. (2016). Bodystorming for movement-based interaction design. *Human Technology, 12*(2), 193–251.

Oulasvirta, A., Kurvinen, E., & Kankainen, T. (2003). Understanding contexts by being there: Case studies in bodystorming. *Personal and Ubiquitous Computing, 7*(2), 125–134.

Oppegaard, B., & Still, B. (2013). Bodystorming with Hawkins's block: Toward a new methodology for mobile media design. *Mobile Media & Communication, 1*(3), 356–372.

Schleicher, D., Jones, P., & Kachur Niedzielski, O. (2010). Bodystorming as embodied designing. *Interactions, 17*(6), 47–51.

Smith, B. (2014). Bodystorming mobile learning experiences. *TechTrends, 58*(1), 71–76.

Recommended Resources

Chang, A. R., & Davis, M. (2005, April). Designing systems that direct human action. In *CHI'05 Extended Abstracts on Human Factors in Computing Systems* (pp. 1260–1263).

Hampshire, N., Califano, G., & Spinks, D. (2022). Bodystorming. In N. Hampshire, G. Califano, and D. Spinks (Eds.), *Mastering collaboration in a product team: 70 techniques to help teams build better products* (pp. 122–123). Apress.

Park, K. W., Yoo, J. W., Park, S. K., Jeong, Y. W., Hwang, W., Park, K. H., . . . Lee, W. (2010). Our experiences on the design and implementation of wearable computers: From body-storming to realized services. In *IEEE International Symposium on ISWC* (pp. 1–3). IEEE.

Schleicher, D., Jones, P., & Kachur, O. (2010). Bodystorming as embodied designing. *interactions, 17*(6), 47–51. https://doi.org/10.1145/1865245.1865256

Segura, E., Vidal, L., & Rostami, A. (2016). Bodystorming for movement-based interaction design. *Human Technology, 12*(2), 193.

See also: Creative Confidence, Embodiment, Context & Contextual Design

85. Card Sorting

Joseph Bartolotta

Part of the user-centered design process is taking an inventory of the user's mental model of how an interactive system operates. While user research methods such as interviews, focus groups, and as-is observations are helpful, card sorting has emerged as another important tool in better understanding user behaviors. Card sorting is instructive to designers as they better understand how to structure the architecture of the interactive system in such a way that users find it intuitive.

At its simplest, card sorting involves giving users cards with different concepts on them and asking them to arrange the cards in structures that make sense to them. Users may be asked to arrange the cards in a sequence, or to group cards by what they consider like terms. This is called "closed card sorting." In some cases, users are also asked to come up with the concepts that will be featured on the cards themselves. This is called "open card sorting."

Here is an example of closed card sorting: Suppose a designer is tasked with redesigning the interface for a band's website. Part of your task is to come up with the structure of the pages within the website. You may suppose that you have the following categories which you will need to include in the website:

News	Upcoming shows	Past Shows	Music Videos	Music
Social Media	Photos	Contact	Band Bio	Store

All of these categories are written on index cards and presented to potential users with the instructions to group like items with each other. As a result, a grouping such as this may appear:

Upcoming Shows Past Shows	Music Videos Streaming Music	Band Bio Photos Social Media
News	Contact	Store

Based on this sorting, the designer may choose to create new categories (such as "Shows" to encompass "Upcoming Shows" and "Past Shows"), and use these new categories to compose a cleaner and more intuitive user journey through the band's website. Having several potential users sort the cards into categories will allow designers to see if patterns emerge towards a generalizable sense of what the user's mental model is when they interact with a system.

Recommended Resources

Barnum, C. M. (2011). *Usability testing essentials: ready, set . . . test!* Morgan Kaufmann. p. 58.

Cooper, A., Reimann, R., Cronin, D., & Noessel, C. (2014). *About face: The essentials of interaction design (4th edition)*. John Wiley & Sons, Inc.

Redish, J. G. (2012). *Letting go of the words: Writing web content that works (2nd edition)* Morgan Kaufmann. pp. 78, 104.

Righi, C., James, J., Beasley, M., Day, D. L., Fox, J. E., Gieber, J., & Ruby, L. (2013). Card sort analysis best practices. *Journal of Usability Studies, 8*(3), 69–89.

Spencer, D. (2009). *Card sorting: Designing usable categories.* Rosenfeld Media.

Wood, J. R., & Wood, L. E. (2008). Card sorting: current practices and beyond. *Journal of Usability Studies, 4*(1), 1–6.

Zimmerman, D. E., & Akerelrea, C. (2002). A group card sorting methodology for developing informational websites. In *Proceedings. IEEE International Professional Communication Conference* (pp. 437–445). IEEE.

See also: Affinity Diagramming, User Experience, User Requirements, User Stories

86. Case Study

Mandy Olejnik

A case study is a method used in qualitative research to closely examine a single participant or small group of participants to learn valuable information about them and their specific contexts ("Case studies," n.d.). Results from a case study are not applicable to everything or to broader cases, given the small and specialized sample size. However, scholars like Anne Haas Dyson (2013) remind us that, with the case study, "its power . . . is that it gives us concrete material with which to think about abstract phenomena, like relevance, agency, and equity" (p. 418). By providing rich, detailed study of a few cases, we can't generalize, but we can point out issues to consider in our broader research, as well as raise questions that help us consider other situations and contexts.

Within technical communication studies and the maker movement, case studies are crucial research approaches used to learn more about how people experience things and how we can further improve or adapt their experiences. As an example in technical communication studies, Elisabeth Kramer-Simpson (2018) conducted a six-month, in-depth case study of three student interns to learn more about the type of mentoring students receive in their internships. She conducted four thirty-minute interviews with each student every one or two months into their internships as well as two interviews with their mentors. Her study was limited to the three students in her study (a small group) who each encountered specific experiences (one student was terminated from their internship due to miscommunication with their supervisor).

Kramer-Simpson's (2018) overall findings in this study include important implications for ways to further support students in internships, such as "it may be helpful to encourage students to identify problems in their writing and why those are issues" (p. 373) and that "students need to take ownership of the concepts discussed in these mentoring conversations" (p. 374). Kramer-Simpson (2018) situates the rich data gained from her case study with other research in technical communication studies, without making generalized claims that this specific context cannot support. She makes recommendations based on these research findings, and based on valuing the students' rich, nuanced experiences. These are experiences that would not have been easily discovered using a non-case study and time-intensive research approach.

Overall, a case study is an important and valuable method to learn about the experiences of individual people or small groups and reflect on what works in similar situations. Many companies also rely on case study research to learn more about their customers and how they experience their products, including Coca-Cola (App Annie, 2018) and *USA Today* (Fantasy, 2018). This demonstrates the broad applicability and strength of the case study as a method, when used and contextualized appropriately.

References

App Annie (2018, March 6). Coca-Cola relies on App Annie to help amaze & delight Its customers [Blog post]. https://www.appannie.com/en/insights/customer-stories/the-coca-cola-company/

"Case Studies." The WAC Clearinghouse. https://wac.colostate.edu/resources/writing/guides/casestudies/.

Dyson, A. H. (2013). The case of missing childhoods: Methodological notes for composing children in writing studies. *Written Communication, 30*(4), 399–427. doi:10.1177/0741088313496383

Kramer-Simpson, E. (2018). Feedback from internship mentors in technical communication internships. *Journal of Technical Writing and Communication, 48*(3), 359–378. doi:10.1177/0047281617728362

Recommended Resources

Bennett, A. (2004). Case study methods: Design, use, and comparative advantages. *Models, numbers, and cases: Methods for studying international relations, 2*(1), 19–55.

Fantasy. (2018). Reimagining the world of news. https://fantasy.co/work/usatoday

Hancock, D. R., Algozzine, B., & Lim, J. H. (2021). *Doing case study research: A practical guide for beginning researchers* (4th ed.). Teachers College Press.

Unterfrauner, E. (2017, May 22). Case study on 10 different maker initiatives.http://make-it.io/2017/05/22/d3–1-case-study-on-10-different-maker-initiatives/

Yin, R. K. (2009). *Case study research: Design and methods.* Sage Publications.

See also: Fieldwork, User Requirements, Local, Environments

87. Diary Study

Elin Björling

Unlike capturing data in a lab, or at a specific set time, a diary study allows for the capture of user data over a period of time, ideally in a natural setting. The real benefit of a diary study is to understand an individual's experience over time. This can be very useful when looking at technology adoption, or to capture accurate, real-time data regarding behaviors, or even moods. Diary studies can capture both quantitative and qualitative data and can be highly customized, or follow a rigid, previously determined instrument depending upon the purpose of the study.

There are two categories of diary studies:

1. End of day diary/log: This type of diary is common in health and behavior research. It's important to remember that highly fluctuating variables like mood and pain are less accurate if only captured retrospectively at the end of the day. However, end of day diaries are great for capturing a user's reflection or summary of the day's events.

2. Within day diary/log: This type of diary is very useful if you want to capture data as it happens. It is very effective for capturing highly fluctuating variables such as mood, pain, or difficult to remember experiences.

Ecological Momentary Assessment (EMA): Similar to the experience sampling method (Csikszent Mihalyi & Larson, 2014) EMA is a common diary approach to capture the experience of users in real time. EMA (Stone & Shiffman, 1994) is characterized by (a) collection of data in real-world environments; (b) assessments that focus on individuals' current or very recent states or behaviors; (c) assessments that may be event-based, time-based, or randomly prompted (depending on the research question); and (d) completion of multiple assessments over time.

Benefits of a diary study: The largest benefit of a diary study, especially using an ecological momentary assessment approach, is the reduction or elimination of recall bias (Shields, Shiffman, & Stone, 2016). Humans are not great at accurate recall, therefore this bias heavily alters self-report data where recall is required. Using a real-time method, ideally capturing data as it happens, can greatly reduce or even eliminate recall bias. In addition, a diary method that captures data *in the wild* also allows for increased ecological validity. This means that the natural environment and context are part of the user's experience and self-report, allowing for increased validity.

Signaling users: An important component of any diary study is reminding the participant to complete a diary. Diary entries can be *signal contingent* (user responds to a signal via a text, or an app randomly or based upon a schedule throughout the day) or *event-contingent* (user creates an entry each time an event occurs). As diaries become increasingly digital, signaling can even be associated with biological events such as heart rate or activity.

Things to watch out for when conducting a diary study: Although diary studies can be paper, internet, or app based, it is important to design your study such that users are reporting data accurately. A common and concerning aspect in diary studies is what is called "hoarding data" where a user forgets to complete diaries and then completes several at one time period (Stone et al., 2003). Therefore, digital diaries provide an ideal environment for data collection as they can be designed to time stamp diary entry data and/or create time-out features to ensure that users are accurately reporting data in real time.

References

Csikszentmihalyi, M., & Larson, R. (2014). Validity and reliability of the experience-sampling method. In M. Csikszentmihalyi (Ed.), *Flow and the foundations of positive psychology* (pp. 35–54). Springer.

Shields, A. L., Shiffman, S., & Stone, A. (2016). Recall bias: Understanding and reducing

bias in PRO data collection. In B. Tiplady and B. Byrom (Eds.), *ePro: Electronic solutions for patient-reported data* (pp. 41–58). Routledge.

Stone A. A., & Shiffman S. (1994). Ecological momentary assessment in behavioral medicine. *Annals of Behavioral Medicine, 16,* 199–202.

Stone, A. A., Shiffman, S., Schwartz, J. E., Broderick, J. E., & Hufford, M. R. (2003). Patient compliance with paper and electronic diaries. *Controlled clinical trials, 24*(2), 182–99.

Recommended Resources

Bolger, N., Davis, A., & Rafaeli, E. (2003). Diary methods: Capturing life as it is lived. *Annual Review of Psychology, 54*(1), 579–616.

Nielsen Norman Group. (n.d.). Diary studies. https://www.nngroup.com/articles/diary-studies/

Shiffman, S. (2016). Ecological momentary assessment. In K. J. Sher (Ed.), *The Oxford handbook of substance use and substance use disorders* (Vol. 2) (pp. 466–509). Oxford University Press.

Shiffman, S., Stone, A. A., & Hufford, M. R. (2008). Ecological momentary assessment. *Annu. Rev. Clin. Psychol.*, *4*, 1–32.

Varese, F., Haddock, G., & Palmier-Claus, J. (2019). Designing and conducting an experience sampling study. In J. Palmier-Claus, G. Haddock, and F. Varese (Eds), *Experience sampling in mental health research* (pp. 8–17). Routledge.

See also: Autoethnography, Ethnography, Memory, User Stories

88. Ethnography

Renee Ann Drouin

S imilar to design thinking, ethnography is a person-focused process dependent on melding diverse practices and thoughts among individuals. Ethnography is a qualitative research method originating from numerous interdisciplinary fields, such as anthropology and sociology (Salkind, 2010; Harrison, 2018). Becky Moss (1992) defines the primary purpose of ethnography is "to gain a comprehensive view of the social interactions, behaviors, and beliefs of a community or social group. In other words, the goal of an ethnographer is to study, explore, and describe a group's culture" (p. 155). Performing ethnographic fieldwork requires both the active participation of the observed "other" and the researcher, gathering data through personal experiences.

Fig. 88.1. Simply by watching how something works, designers can understand user needs and requirements. Photo by Remy Gieling on Unsplash. https://bit.ly/keyword-fig-88–1

Methods of ethnographic fieldwork prioritize observation, interviews, and archival analysis (Salkind, 2010). Observation stems from the researcher's firsthand experience in witnessing the community interactions. Community members are interviewed or surveyed with tools designed and made specifically for their context and community by the researcher. Members or individuals may provide artifacts for examination. Though understanding the community is a priority, scholars similarly recognize the benefits of the researcher's self-reflexivity, the act of locating and defining their presence within the research, regardless of if they consider themselves a member of the community (Chiseri-Strater, 1996; Sullivan, 1996).

Means of making and designing tools for ethnography are dependent on the space and location of the documented community. One of the earliest challenges ethnographers face is determining place and time, as the "spatial boundaries" are more complex to define without a physical space and the multi-sites of the internet. Historically, fieldwork has been physical in nature, though the development of digital spaces have led to divergent methods and terms. Online ethnography, cyber-ethnography, virtual ethnography, and *net*nography are common terms of the networked ethnographic studies (Steinmetz, 2012; Costello, McDermott, & Wallace, 2017). Unique dilemmas of ethnographies performed online concern ages of consent, anonymity, and positioning of the researcher, and how ethnographic materials are made and designed.

Despite the considerations of location, community, and design that seemingly create structured parameters of how to conduct research, improvisation is vital to performing ethnography (Harrison, 2018). Fieldwork is a social dynamic that requires ethnographers to be "continually adjusting to the information they receive and the experiences they have. In this sense they must be spontaneous strategists" (p. 44). Performing ethnography through such a mindset emphasizes the traits of design thinking and how they apply to fieldwork. Nadia M. Viljoen and Ria Van Zyl (2009) attribute inventiveness, collaborative teamwork, optimism, and an ability to communicate to design thinkers, facets extended towards the improvisation and social dynamics of ethnography.

Carrie Leverenz's (2014) observation of ethnography in designing shows the social dynamic in divergent interests of actors in design (materials, users, etc.) coming together. An example of this (Cross, 2011) positions designers as cognizant of the collaboration required and how different needs must be met.

287

References

Chiseri-Strater, E. (1996). Turning in upon ourselves: Positionality, subjectivity, and reflexivity in case study and ethnographic research. In P.

Mortensen and G. Kirsch (Eds.), *Ethics and representation in qualitative studies of literacy* (pp. 115–133). National Council of Teachers of English.

Costello, L., McDermott, M. L., & Wallace, R. (2017). Netnography: Range of practices, misperceptions, and missed opportunities. *International Journal of Qualitative Methods*, *16*, 1–12.

Cross, N. (2011). *Design thinking: Understanding how designers think and work.* Bloomsbury Academic.

Harrison, A. K. (2018). *Ethnography*. Oxford University Press.

Moss, B. (1992). Ethnography and composition: Studying language at home. In G. Kirsch and P.A. Sullivan (Eds.), *Methods and methodology in composition research* (pp. 153–171). Southern Illinois University Press.

Salkind, N. J. (2010). *Encyclopedia of research design*. Sage Publications.

Steinmetz, K. F. (2012). Message received: Virtual ethnography in online message boards. *International Journal of Qualitative Methods*, *11*(1), 26–39.

Sullivan, P. (1996). Ethnography and the problem of the "Other." In P. Mortensen and G. Kirsch (Eds.), *Ethics and representation in qualitative studies of literacy* (pp. 97–114). National Council of Teachers of English.

Viljoen, N. M., & Van Zyl, R. H. M. (2009). Design thinking—crossing disciplinary borders. *Image & Text: A Journal for Design*, *15*, 66–78.

Recommended Resources

Kimbell, L. (2011). Rethinking design thinking: Part I. *Design and Culture*, *3*, 285–306.

Kimbell, L. (2012). Rethinking design thinking: Part II. *Design and Culture*, *4*(2), 129–148.

LaFave, A. L., & Mainz, E. A. (2018). Engaging with the syuzhet: A new methodological approach to analyzing and visualizing internet discourse. *Ethnography*, *19*(2), 183–203.

Leverenz, C. (2014). Design thinking and the wicked problem of teaching writing. *Computers and Composition*, *33*, 1–12. doi:10.1016/J.COMPCOM.2014.07.001

Vindrola-Padros, C. (2021). *Rapid ethnographies: A practical guide*. Cambridge University Press.

See also: Autoethnography, Fieldwork, User Experience, User Stories

89. Fieldwork

Casey Boyle

Fieldwork in the social sciences describes a wide array of techniques for gathering qualitative research. These techniques are multiple and can refer to ethnography, interviews, surveys, direct and indirect observation, document analysis and, increasingly, techniques involving computation and audio/visual media. Some disciplines that rely on fieldwork include (but are not limited to) anthropology, geology, sociology, psychology, user experience, design, political science and, more recently, rhetorical studies. As with any term that gets deployed across such a wide-ranging set of disciplines, the concept is always and already multiple but it also coheres as a quasi-singular idea. There are key tensions in the keyword of field work that must be considered including its colonialist origins, research ethics, delineations of a field, and the future of fieldwork as digital and communication technologies topologically reshape fields into streams.

Like most techniques for producing knowledge in and for the Western tradition, fieldwork has had a vexed history. Its earliest practices begin with colonialism whereby the learned gentleman [sic] would travel to and inhabit a "field" to observe customs of "natives" and report back through writing and scholarship. This history both continues as much as it is challenged and being reshaped. Shannon Mattern (2016) writes that "the contemporary fetishization of ethnography, and the consequent diffusion (and popularization) of fieldwork, the method is far from pure. It's wrapped up in colonialist and gendered ideologies" (n.p). Indeed, while Mattern is a media theorist, disciplines such as anthropology, sociology, and psychology have for decades been wrestling with their disciplinary ties to colonialist knowledge practices. Such intrusions by privileged researchers into the habits and daily practices of (all too often) a less-privileged community bring up a host of troubling issues that should give an ethical research pause if not full stop. One of the best examples that articulates these tensions in fieldwork may be Paul Rabinow's (2007) questioning of the validity of the fieldwork process paying special attention to the relationship between the ethnographer/field workers and their informant.

In addition to questioning the practices of fieldwork that fetishize and exoticize an other, social sciences also has begun to challenge what even constitutes a field in which one may perform researched work. Doreen

289

Massey (2003) proposes that the field itself is blurry, always co-produced by assumptions and connections that start well before and extend far after any individual self chooses to enter a "field." Michelle Kisliuk (1997) similarly questions a field by turning to the experience of music and sound as one that cannot be located outside an observer in whom the experience of a field is embodied. What these works suggest, in varying implicit and explicit frequencies, is how a field might itself be leaky and not well defined, especially in an increasingly globally connected world.

As the fields we might research change and become more connected through information and communication technologies, so do our methods of performing field change as those same technologies introduce new and innovative ways to perform fieldwork. For instance, fieldwork becomes radically expansive in the era of globalization as shown by *Unknown Fields* (http://www.unknownfieldsdivision.com/), an ongoing research and design project that gathers multidisciplinary researchers to explore energy fields, electronic mineral extraction, fashion production and several other sites. This project uses multiple disciplines and technologies to document—writing, designing, filming, etc.—the wide array of practices that go into any *one* field of study. Related, if slightly less radical projects, can understand the fieldwork as tracing how events of protest can be and are extended through social networking. Work by Zizi Papachrissi (2015), Zeynep Tufekci (2017), and Jennifer Gabrys (2016) explore how the digital is reshaping our fields of study and, with them, our own understanding of who we are and what we are capable of.

References

Gabrys, J. (2016). *Program earth: Environmental sensing technology and the making of a computational planet.* University of Minnesota Press.

Kisliuk, M. (1997). (Un)doing fieldwork: Sharing songs, sharing lives. In G. F. Barz and T. J. Cooley (Eds.), *Shadows in the field: New perspectives for fieldwork in ethnomusicology* (pp. 23–44). Oxford University Press.

Massey, D. (2003). Imagining the field. In M. Pryke, G. Rose, and S. Whatmore (Eds.), *Using social theory* (pp. 71–88). Sage Publications.

Mattern, S. (2016). Cloud and field. *Places Journal.* https://placesjournal.org/article/cloud-and-field/?cn-reloaded=1

Papacharissi, Z. (2015). *Affective publics: Sentiment, technology, and politics.* Oxford University Press.

Rabinow, P. (2007). *Reflections on fieldwork in Morocco.* University of California Press.

Tufekci, Z. (2017). *Twitter and tear gas: The power and fragility of networked protest.* Yale University Press.

Recommended Resources

Conquergood, D. (1992). Ethnography, rhetoric, and performance. *Quarterly Journal of Speech, 78*(1), 80–97.

Endres, D., Hess, A., Senda-Cook, S., & Middleton, M. K. (2016). In situ rhetoric: Intersections between qualitative inquiry, fieldwork, and rhetoric. *Cultural Studies Critical Methodologies, 16*(6), 511–524.

Goffman, E. (1989). On fieldwork. *Journal of Contemporary Ethnography, 18*(2), 123–132.

Rai, C., & Druschke, C. G. (Eds.). (2018). *Field rhetoric: Ethnography, ecology, and engagement in the places of persuasion.* University of Alabama Press.

Young, L., & Davies, K. (2013). A distributed ground: The unknown fields division. *Architectural Design, 83*(4), 38–45.

See also: Case Study, Interviewing, Ethnography, Autoethnography

90. Interviewing

Ashanka Kumari

Interviewing has long served as a qualitative method for gathering information about people. For designers, interviewing operates as a technique for learning more about the population(s) that will engage with the end design. Additionally, interviewing serves as one part of testing a design before bringing it forward to a larger population.

Fig. 90.1. Conversations between stakeholders can inform the design of any process. Photo by National Cancer Institute on Unsplash. https://bit.ly/keyword-fig-90-1

To demonstrate the value of interviewing for design thinkers and makers, consider its use in a Biomedical Engineering Design course at the University of Texas at Austin (Markey, Monteiro, & Stewart, 2018). In this class, students explored "the design of health information systems for supporting medical decision-making." To do so, students first prepared to do semi-structured interviews using techniques they learned from relevant course readings and discussions. Following this, students conducted interviews with healthcare professionals toward determining actionable problems for which they could design solutions. As demonstrated in this case, it is crucial to interview the population for which you are hoping to design something prior to designing it in order to best serve their specific needs.

Alongside semi-structured interviewing strategies, some designers have also used *empathy interviews* as a method toward problem-solving for different communities. Stacey Messier (2017) describes an empathy interview as one where a moderator (question-asker) and note-taker are appointed and questions are asked for thirty minutes to an hour either in-person, over the phone, or video chat to a target design customer. The moderator focuses on asking open-ended questions and listening rather than talking.

Empathy interviews offer potential customers or users of designs to relay what they feel is most important to them with a "focus on the emotional or subconscious aspects of the user" (Sa, 2018). For interviewers, this approach allows an opportunity to "observe body language and reactions of the subjects," and build additional follow-up questions based on their observations (Sa, 2018). In preparation for any kind of interview, designers might conduct an "assumptions workshop" during which they gather design team members and any stakeholders and talk for about an hour about the following questions: What do we know? What do we want to know? What do we assume? This workshop is designed to help designers further determine the target consumer of their projected design. Further, this workshop can aid in building questions for moderators to later use in empathy interviews (Messier, 2017).

Designers can also engage interviewing strategies after spending some time creating their design after initial interviews. This additional round of interviewing can offer designers insight into whether what they have designed works and make changes as necessary. Here, designers should give community members an opportunity to engage with the design and then ask open-ended questions from which designers can reflect and continue to refine the design for greater community use. Finally, designers should return after some time and interview the community again to ensure that the design continues to serve its purpose. Overall, interviewing should occur throughout the design thinking process to better understand the needs of the population for which the design is intended.

References

Markey, M. K., Monteiro, J. C., & Stewart, J. (2018). Using Twitter to support students' design thinking. In *Proceedings of the 2018 ASEE*

Gulf–Southwest Section Annual Conference, University of Texas at Austin, April 4–6, 2018.

Messier, S. (2017, Jan. 24). Design thinking: What is an empathy interview? *Medium.* https://medium.com/@StaceyDyer/design-thinking-what-is-an-empathy-interview-25f71bd496d7

Sa, L. L. (2018, June 26). Techniques for empathy interviews in design thinking. https://webdesign.tutsplus.com/articles/techniques-of-empathy-interviews-in-design-thinking—cms-31219

Recommended Resources

Bevan, M. T. (2014). A method of phenomenological interviewing. *Qualitative Health Research, 24*(1), 136–144.

Döringer, S. (2021). The problem-centred expert interview: Combining qualitative interviewing approaches for investigating implicit expert knowledge. *International Journal of Social Research Methodology,* 24(3), 265–278.

Liedtka, J. (2018). Exploring the impact of design thinking in action. *Darden working paper series.* Retrieved from https://designatdarden.org/app/uploads/2018/01/Working-paper-Liedtka-Evaluating-the-Impact-of-Design-Thinking.pdf

Micheli, P., Wilner, S. J., Bhatti, S. H., Mura, M., & Beverland, M. B. (2019). Doing design thinking: Conceptual review, synthesis, and research agenda. *Journal of Product Innovation Management, 36*(2), 124–148.

Nardon, L., Hari, A., & Aarma, K. (2021). Reflective interviewing—Increasing social impact through research. *International Journal of Qualitative Methods,* 20, 1–12. https://doi.org/10.1177/16094069211065233

See also: User Stories, Journey Mapping, Card Sorting

91. Journey Mapping

Joe Moses

A journey map visualizes a process that people follow in order to accomplish a goal (Gibbons, 2018). A good journey map represents a research process and the findings that follow from contact with real people. Used in design thinking to increase developer empathy, journey maps—also called customer journey maps—display ways in which users interact with information during their decision-making processes.

For commercial purposes, developers base journey maps on data that reflect user preferences or questions and—especially—obstacles that users describe when browsing websites or social media. The research process typically results in a variety of findings about customers:

- Customer needs and interests
- Obstacles customers face when searching for information
- Ideas for improving user experience

As a tool for describing a process and wayfinding, journey mapping can serve a variety of non-commercial purposes as well. Journey maps consist of 1) a person, 2) scenarios that describe a need for a purpose, 3) phases of the journey, 4) attitudes and behaviors, and 5) insights.

Person. Interviews with real people should result in a profile that captures actual user interests or composites of common user interests. Profiles typically include specifics about—

1. How much experience the user has with a process, product, or service.
2. How the user prefers to learn new information or skills about a process, product, or service.
3. How the user typically goes about problem solving.

Scenarios. The classic model for user scenarios gives users a chance to imagine a person in a role with a need for a purpose:

Role: As a frequent flyer,
Need: I want an easy way to book flights,
Purpose: so I can find good deals quickly.

Phases of the Journey. Most people have more than one need, so journey maps show where in a process those needs and purposes arise. By showing intersections of user needs with steps in a process, journey maps indicate where and why obstacles occur and help suggest how to improve usability.

Attitudes and Behaviors. Some of the value of journey maps comes from their intersections of time with points of decision-making, which helps analysts understand what a customer is experiencing at different phases of a process. The following example shows the intersection of a user's thinking with two phases of their journey: researching and booking a flight. The sample includes the user's expression of attitudes—"great customer reviews," and "looks like it will slow me down" and a behavior: "I'm going to skip it."

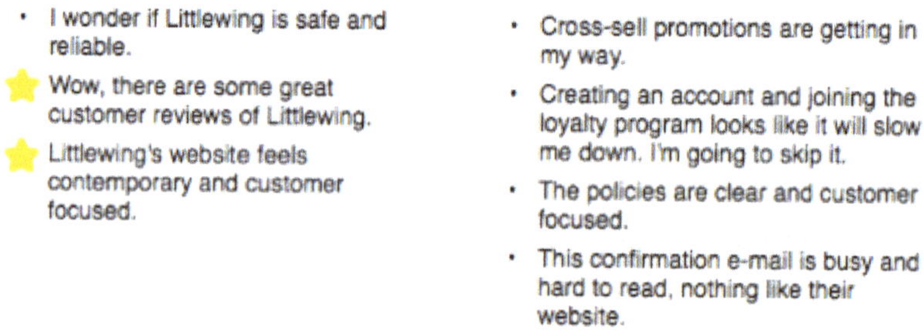

- I wonder if Littlewing is safe and reliable.
- ⭐ Wow, there are some great customer reviews of Littlewing.
- ⭐ Littlewing's website feels contemporary and customer focused.

- Cross-sell promotions are getting in my way.
- Creating an account and joining the loyalty program looks like it will slow me down. I'm going to skip it.
- The policies are clear and customer focused.
- This confirmation e-mail is busy and hard to read, nothing like their website.

Fig. 91.1. Intersection of a user's thinking with two phases of their journey reveals user attitudes and behaviors while searching a website (Davey, 2018).

Insights. Based on the data gathered from a frequent business traveler, above, creators of a journey map are able to derive insights from the scenarios, phases, attitudes, and behaviors they captured. If you go to the larger image (Davey, 2018, https://bit.ly/keyword-journey-map), you will see a list of opportunities suggested by the journey map, which comprise insights about the data, including ways of increasing customer awareness and brand loyalty.

Productive Journey Mapping. Given the apparent simplicity of well-researched journey maps and the value of information in them, organizations and individuals often mistakenly start with existing data about others' customers instead of meeting with their own customers. While data about others' customers can provide some valid information, interacting with real people instead of someone else's data is more timely and three-dimensional.

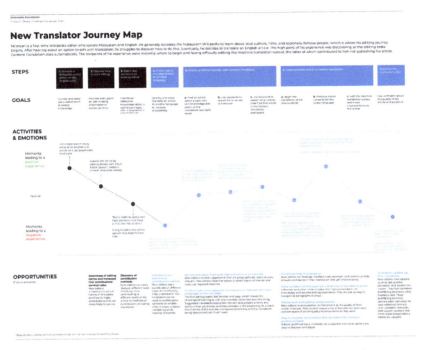

Fig. 91.2. A sample journey map (textual content not important) for a new technical translator. Graphic by EAsikingarmager (WMF) via Wikimedia Commons. https://commons.wikimedia.org/wiki/File:New_Translator_Journey_Map.png

Empathy. Journey mapping can help analysts understand how other people feel while engaging with their organizations. While empathy is important, it's not an end in itself. Design thinkers pursue empathy so they can accurately define problems to address on behalf of their organization and its interests. In short, journey mapping is as much a journey for those who set out to research their customers as it is about understanding the customer's journey.

References

Davey, N. (2018). Nine sample customer journey maps—and what we can learn from them. MyCustomer.com. https://www.mycustomer.com/experience/engagement/nine-sample-customer-journey-maps-and-what-we-can-learn-from-them

Gibbons, S. (2018). Journey mapping 101. Nielsen Norman Group. https://www.nngroup.com/articles/journey-mapping-101/

Recommended Resources

Bradley, C., Oliveira, L., Birrell, S., & Cain, R. (2021). A new perspective on personas and customer journey maps: Proposing systemic UX. *International Journal of Human-Computer Studies, 148,* 102583. https://doi.org/10.1016/j.ijhcs.2021.102583

Heuchert, M. (2019). Conceptual modeling meets customer journey mapping: Structuring a tool for service innovation. In *2019 IEEE 21st Conference on Business Informatics* (pp. 531–540). IEEE.

Ludwiczak, A. (2021). Using customer journey mapping to improve public services: A critical analysis of the literature. *Management, 25*(2), 22–35. https://doi.org/10.2478/manment-2019–0071

Micheaux, A., & Bosio, B. (2019). Customer journey mapping as a new way to teach data-driven marketing as a service. *Journal of Marketing Education, 41*(2), 127–140. https://doi.org/10.1177/0273475318812551

Tueanrat, Y., Papagiannidis, S., & Alamanos, E. (2021). Going on a journey: A review of the customer journey literature. *Journal of Business Research, 125,* 336–353. https://doi.org/10.1016/j.jbusres.2020.12.028

See also: User Requirements, User Stories, Context & Contextual Design

92. Peer Response

Heather Listhartke

Also known as peer review, peer evaluation, or peer editing, peer response is the act of working with a peer to give and gain feedback on content development and style of writing. The peers may offer suggestions, questions, corrections, reactions, and support to each other's work of writing shared.

The term was originally used in the late 1970s with one of the first publications being an article published in 1979 by the *Language Arts* journal called "The Writing Workshop: An experiment in peer response to writing" written by Marion Crowhurst. She calls peer response the solution to teacher overload on "marking" when students need to know more about writing by performing more writing (Crowhurst, 1979, p. 757).

That same year, Nancy Benson (1979), in her dissertation at the University of Colorado entitled "The effects of peer feedback during the writing process on writing performance, revision behavior, and attitude toward writing" did a study, which was heavily cited for much of the next decade pointing to improvements in student revisions.

Currently, peer response is usually built into the student outcomes of the writing and communication classroom, especially in first year writing. Many teachers work it into all stages of the writing process with major peer response happening when students have a mostly complete draft, often referred to as a zero draft or a "shitty first draft" in alluding to Anne Lamott's (2005) article of the same name. This change in the way that peer revision has been looked at and used in the last forty years is quite significant as the method isn't about reducing instructor marks as much as it is about the positive gains in writing skills and understanding of the writing process that we know students have when responding to and getting responses from their peers. It's also that as a field we've come to understand thanks to extensive work by Linda Adler-Kassner and Elizabeth Wardle (2016) that "writing is social" and peer response is one way of helping students see that.

Much of the current scholarship on peer response is not only about how to get students to respond more effectively to their peers and get them to more effectively implement feedback they receive, but it is also trying to find the best methods and pedagogies to work with English language learners as they continue to develop their writing. Further still, one of the key places where effective strategies and pedagogies are being researched in the field is the work that writing centers do with students. While their inclusion in colleges and now high schools continues to grow, they are increasingly finding the added benefit of responding to peers' work on one's own writing. It's interesting to note that there has even been the development of online guides for peer responses like the Eli Review tool (https://elireview.com), which originally started at Michigan State University writing programs.

One of the most popular methods of peer review still remains in the writing classroom in which students are split into small groups or pairs, exchange nearly complete drafts, and then respond to each other's writing. However, the things that happen in peer response have continued to change. The focus of peer response is no longer to read like an editor correcting grammar and mechanics, but more about the development of the content being asked for, the logical arrangement of the argument, and the attention to rhetorical and genre awareness. In short, peer response asks that peers respond to the works as readers and audience members as Peter Straub (1999) calls for in his article "Responding—Really Responding—to Other Students' Writing."

References

Adler-Kassner, L., & Wardle, E. A. (2016). *Naming what we know: Threshold concepts of writing studies*. Utah State University Press.

Benson, N. (1979). *The effects of peer feedback during the writing process on writing performance, revision behavior, and attitude toward writing*. Doctoral dissertation. University of Colorado.

Crowhurst, M. (1979). The writing workshop: An experiment in peer response to writing. *Language Arts, 56*(7), 757–762.

Lamott, A. (2005). Shitty first drafts. In P. Eschholz, A. Rosa, and V. Clark (Eds.), *Language awareness: Readings for college writers* (9th ed., pp. 93–96). Bedford St. Martin's.

Straub, P. (1999). Responding—really responding—to other students' writing. In *The subject is writing* (2nd ed., pp. 136–146). Boynton/Cook.

Recommended Resources

Anvari, F., Richards, D., Hitchens, M., & Tran, H. M. T. (2019, May). Teaching user-centered conceptual design using cross-cultural personas and peer reviews for a large cohort of students. In *2019 IEEE/ACM 41st International Conference on Software Engineering: Software Engineering Education and Training* (pp. 62–73). IEEE. https://ieeexplore.ieee.org/abstract/document/8802096

Bunn, M. (2010). How to read like a writer. In *Writing spaces* (Ser. 2). https://writingspaces.org/bunn—how-to-read-like-a-writer

Demmans Epp, C., Akcayir, G., & Phirangee, K. (2019). Think twice: exploring the effect of reflective practices with peer review on reflective writing and writing quality in computer-science education. *Reflective Practice, 20*(4), 533–547. https://doi.org/10.1080/14623943.2019.1642189

Panadero, E., & Alqassab, M. (2019). An empirical review of anonymity effects in peer assessment, peer feedback, peer review, peer evaluation and peer grading. *Assessment & Evaluation in Higher Education, 44*(8), 1253–1278. https://doi.org/10.1080/02602938.2019.1600186

Reddy, K., Harland, T., Wass, R., & Wald, N. (2021). Student peer review as a process of knowledge creation through dialogue. *Higher Education Research & Development, 40*(4), 825–837. https://doi.org/10.1080/07294360.2020.1781797

See also: Active Learning, Social Constructivism, Mentoring

93. Photovoice

Erin Brock Carlson

Photovoice, also known as participant-generated imagery (PGI), is a participatory action research method that initially emerged from public health and public policy (Wang & Burris, 1997) and has been recently introduced to technical communication (Sullivan, 2017). In their simplest form, photovoice projects ask participants to take photos of their daily lives and then to reflect on their photos in order to find connections, patterns, and relationships that otherwise might remain hidden. Drawing on Paulo Freire's critical consciousness and liberatory ideology (1970), photovoice aims to disrupt power dynamics inherent in traditional research; that is, it positions participants as researchers who manage the generation, collection, and interpretation of data themselves. Placing control in participants' hands (or, cameras) gives participants a platform for knowledge creation that can be incredibly fruitful in design thinking contexts that require ingenuity and creativity.

Fig. 93.1. Photos generated by participants provided additional contexts to understand research problems. Photo by Aneta Pawlik on Unsplash. https://bit.ly/keyword-fig-93–1

In projects that use this method, participants take photos on a regular basis over a period of weeks or months, often guided by a general prompt. Then, in interviews or focus groups, participants interpret their photos, identifying patterns across their images. During focus groups, participants are asked questions about their photos (i.e., "Which of your photos is the most/least important?" or "Can you group or organize your photos according to theme?") so that they can work collaboratively to understand the similarities and differences between their experiences. Interviews, on the other hand, offer individual participants opportunities to delve more deeply into their own interpretations of the photos and the contexts that shaped their experiences. In addition to interviews, researchers often gather other types of data to supplement pictures. Sometimes, this supplementary data takes the form of written or spoken narratives gathered during the same time period as the photographs; other times, researchers ask participants to curate their photographs and to write reflections about their experiences after the project has ended.

Because photovoice projects rest upon the decisions made by participants, these projects provide participants with a lot of freedom in how they approach a project. This freedom meshes with the following goals of the method, as outlined by Caroline C. Wang (1999):

1. To enable people to record and reflect their community's strengths and concerns;

2. To promote critical dialogue and knowledge about important community issues through large and small group discussions of photographs; and

3. To reach policymakers and those who hold the power to create change.

At its core, photovoice methods are inherently human-centered and undergirded by the belief that the ways a problem is depicted and articulated shape the solutions that emerge.

Photovoice has been used to study a variety of different issues, including health and wellness, environmental changes, urban planning, and community organizing. Photovoice operates like an autoethnography, as participants build their own narrative data by making rhetorical choices as they document their experiences. This makes it especially useful for groups that are disenfranchised or marginalized by institutional powers, or those dealing with difficult circumstances that can be isolating. In Joyce P. Yi-Frazier et al.'s (2015) study on public health and diabetes, participants took pictures of moments they felt captured what it was like to live with diabetes. They then uploaded their photos to Instagram and tagged them with specific hashtags linked to the study. As a result, researchers could easily access photo data, and participants were able to interact with each other through Instagram. During focus groups, researchers found that crafting social networks through digital platforms were integral to participants' well-being while managing a chronic disease—a finding that presumably wasn't on the radar of most healthcare specialists. While this project undoubtedly highlighted some anticipated findings about life with diabetes, its participatory nature guided researchers to unexpected strategies for managing life with diabetes that had an immediate impact on participants.

Photovoice, then, might contribute two major strengths to design thinking. First, given the method's flexibility, it can be used to study experiences and phenomena that are subjective in nature. Second, it allows participants to produce valuable descriptive data that highlights contextual factors of problems. As a result, photos are not just data points, but doorways for participants and researchers, inviting them to examine the complex relationships surrounding each shot.

References

Sullivan, P. (2017). Participating with pictures: Promises and challenges of using images as a technique in technical communication research. *Journal of Technical Writing and Communication, 47*(1), 86–108.

Wang, C. C. (1999). Photovoice: A participatory action research strategy applied to women's health. *Journal of Women's Health, 8*(2), 185–92. DOI: 10.1089/jwh.1999.8.185

Wang, C., & Burris, M. (1997). Photovoice: Concept, methodology, and use for participatory needs assessment. *Health Education & Behavior, 24*(3), 369–387. DOI: 10.1177/109019819702400309

Yi-Frazier, J. P., Cochrane, K., Mitrovich, C., Pascual, M., Buscaino, E., Eaton, L., Panlasigui, N., Clopp, B., & Malik, F. (2015). Using Instagram as a modified application of photovoice for storytelling and sharing in adolescents with type 1 diabetes. *Qualitative Health Research, 25*(10), 1372–1382. https://doi.org/10.1177/1049732315583282

Recommended Resources

Bell, S. E. (2008). Photovoice as a strategy for community organizing in the Central Appalachian coalfields. *Journal of Appalachian Studies, 14*(1/2), 34–48. http://www.jstor.org/41446801

Brock Carlson, E., & Caretta, M. A. (2023). Collaborative sensemaking through photos: Using photovoice to study gas pipeline development in Appalachia. *Qualitative Research*. Online First. https://doi.org/10.1177/14687941221149582

Brock Carlson, E., & Overmyer, T. (2018). Smart phones and photovoice: Exploring participant lives with photos of the everyday. In L. Levenberg, T. Neilson, and D. Rheams (Eds.), *Research methods for the digital humanities* (pp. 129–150). Palgrave Macmillan.

Downey, L., Ireson, C., & Scutchfield, F. (2009). The use of photovoice as a method of facilitating deliberation. *Health Promotion Practice, 10*(3), 419–427. https://doi.org/10.1177/1524839907301408

Freire, P. (1970). *Pedagogy of the oppressed*. Trans. M.B. Ramos. Continuum.

Harper, D. (2002). Talking about pictures: A case for photo-elicitation. *Visual Studies, 17*(1), 13–26. https://doi.org/10.1080/14725860220137345

Harrison, B. (2004). Snap happy: Toward a sociology of "everyday" photography. *Studies in Qualitative Methodology 7*, 23–39. http://dx.doi.org/10.1016/s1042-3192(04)07003-x

See also: Materiality, Visual Rhetoric, Visual Semiotics, Memory

94. Postmortem

Kristopher Purzycki

Compiled at the completion of a project, the postmortem is a genre of formal documentation that provides project managers with an opportunity to assess and reflect on the mistakes and successes that were significant. Although successful practices are often included, postmortems traditionally focus on the obstacles encountered and mistakes committed during the timeline of the project. The purpose of this attention to human errors in calculation and judgment is not to address issues that hindered the completion of a project but to analyze those issues to develop more efficient processes (Collier et al., 1996). In this way the postmortem, or "post-project" becomes integrated into subsequent projects (Disterer, 2002).

Adopted as a practice by computer scientists, the postmortem was considered vital to the recursive quality of software development. Yet the process was amorphous and poorly defined until Bonnie Collier, Tom De Marco, and Peter Fearey (1996) set out to establish a criteria based on team surveys as well as the collection of data pertaining to costs, scheduling, and quality. With metrics created from both subjective and objective sources, teams are then debriefed on the findings. This is followed by an extensive session, what Collier et al. (1996) refer to as the "Project History Day" (p. 69), where participants investigate the root causes of the biggest issues impacting the project.

While postmortems are primarily intended for internal use, media developers and publishers have increasingly released public-facing postmortems. Technical articles, with their formal austerity, are still typical of the genre and are readily available online. More accessible postmortems, however, are emerging as a popular genre published by major media producers such as Facebook and Netflix. The computer game industry, with its roots in engineering laboratories, has been among the chief adopters of the postmortem and continues to embrace the genre as a vital component of the design process. The Game Developers Conference (GDC), for example, regularly hosts designers of classic video games which share postmortems of popular titles. Noted designer Raph Koster's (2017) compilation of "self-evaluative writings and lessons learned," is also appropriately titled *Postmortems*.

As a form of reflective communication, the postmortem has also taken root in technical communications courses, especially those using computer games as curricular vehicles (Ritter et al., 2014). Not only does the postmortem provide a structure for reflection but also a model for professionalization. Incorporated into the design classroom, the postmortem provides a capstone project for the end of the semester with adequate provision. Rather than an afterthought, however, the postmortem should be treated as an integral part of the project that offers a springboard for later iterations rather than as a reflection that signifies completion.

References

Collier, B., Demarco, T., & Fearey, P. (1996). A defined process for project post mortem review. *IEEE Software, 13*(4), 65–72.

Disterer, G. (2002). Management of project knowledge and experiences. *Journal of Knowledge Management, 6*(5), 512–520.

Koster, R. (2017). *Postmortems: Selected essays volume one.* Altered Tuning Press.

Ritter, C., Ansari, S., Daner, S., Murray, S., & Reeves, R. (2014). From realism to reality: A postmortem of a game design project in a client-based technical communication course. In J. Winter and R. M. Moeller (Eds.), *Computer games and technical communication: Critical methods and applications at the intersection* (pp. 283–306). Routledge.

Recommended Resources

Bardzell, J., Bardzell, S., Dalsgaard, P., Gross, S., & Halskov, K. (2016). Documenting the research through design process. In *Proceedings of the 2016 ACM Conference on Designing Interactive Systems* (96–107). ACM. https://doi.org/10.1145/2901790.2901859

Cheng, A., Eppich, W., Grant, V., Sherbino, J., Zendejas, B., & Cook, D. A. (2014). Debriefing for technology enhanced simulation: A systematic review and meta analysis. *Medical Education, 48*(7), 657–666.

Mirza-Babaei, P., Stahlke, S., Wallner, G., & Nova, A. (2020). A postmortem on playtesting: Exploring the impact of playtesting on the critical reception of video games. In *Proceedings of the 2020 CHI Conference on Human Factors in Computing Systems* (pp. 1–12). ACM.

Raemer, D., Anderson, M., Cheng, A., Fanning, R., Nadkarni, V., & Savoldelli, G. (2011). Research regarding debriefing as part of the learning process. Simulation in *Healthcare, 6*(7), S52–S57.

Taylor, J. L., Soro, A., Roe, P., Lee Hong, A., & Brereton, M. (2018, April). "Debrief o'clock": Planning, recording, and making sense of a day in the field in design research. In *Proceedings of the 2018 CHI Conference on Human Factors in Computing Systems* (pp. 1–14). ACM. https://doi.org/10.1145/3173574.3173882

See also: Reflection, Memory, Feedback, Heuristics

95. Reflection

Kathleen Blake Yancey

Several scholars have provided definitions for reflection. Educational theorist Carole Rodgers (2002), borrowing from John Dewey, identifies four defining features of reflection: it is, she says, a "meaning-making process"; a "systematic, rigorous, disciplined way of thinking"; a community-based practice; and a practice requiring "attitudes that value the personal" (p. 845). Business professors Giada Di Stefano, Francesca Gino, Gary Pisano, and Bradley Staats, writing in a 2014 Harvard Business Working Paper, define reflection "as the intentional attempt to synthesize, abstract, and articulate the key lessons learned from experience" (p. 3). Focusing on writing, learning, and portfolios, Kathleen Blake Yancey defines reflection in three ways: first, as an opportunity to make connections, and meaning, from the past, often as a stage for the future (Yancey, 2019); second, as a process by which one can assess his or her own work, which allows one to make connections, think about learning, and to make knowledge (Yancey, 1998); and third, as synthetic, attentive, agentive, juxtapositional, social, and assembled (Yancey, 2016b).

In rhetoric and composition, there have been three generations of work on reflection (Yancey, 2016a). The first generation identified reflection as part of the writing process (Pianko, 1979), then more specifically as linked to revision (Perl, 1980), two associations it retains. The second generation of reflection, beginning in about 1990, employed reflection in assessment, especially portfolios (Neal, 2016) and Directed Self Placement (e.g., San Jose State University's "Reflection on College Writing FAQ"). The third generation of reflection, circa 2000-present, includes reflection and the digital (Silver, 2016); reflection and electronic portfolios (Clark, 2016); and reflection and transfer of writing knowledge and practice (Yancey, Robertson, & Taczak, 2014).

As a construct, reflection is related to revision, but is separate from it (Lindemann et al., 2018). Reflection is also related to, but not synonymous with, metacognition: the latter focuses on self-monitoring and self-assessment, while reflection provides a mechanism for the making of meaning and of knowledge (Yancey, 2016 a). Reflection is also culturally influenced (Inoue & Richmond, 2016), especially along engagements related to gender, ethnicity, and class.

In educational settings involving writing, meaning is made through reflection by students in the classroom (e.g., Sommers, 2016); by faculty developing WAC (writing across the curriculum) and WID (writing in the disciplines) curricula (e.g., Flash, 2016); and by researchers (e.g., Roozen, 2016; Sommers, 2011). Faculty in several disciplines employ reflection-for-learning: for instance, research shows that students who write reflective summaries in introductory physics perform better on tests than those who write only summaries (Kalman et al., 2015). Likewise, the National Science Foundation funded the Consortium to Promote Reflection in Engineering Education (CREE, 2014–2017) to provide guidance for faculty in a range of STEM fields regarding reflective assignments and exercises. More recently, several institutions have redesigned their general education programs to include capstones culminating in electronic portfolios and accompanying reflections: students at Thomas Jefferson University use reflection to speak to their identities as learners, citizens, and future professionals (Schrand et al., 2018), while students at the University of Albany use reflection to document and explain failure; to connect out-of-school learning to in-school learning; and to articulate digital citizenship (Emerson & Reid, 2019). Reflection provides for meaning-making outside of educational settings as well, including in medicine (Gawande, 2003) and architecture (Schon, 1987).

References

Clark, J. E. (2016). From selfies to self-representation in electronically mediated reflection: The evolving gestalt effect in ePortfolios. In K. B. Yancey (Ed.), *A rhetoric of reflection* (pp. 149–165). Utah State University Press.

CPREE/Consortium to Promote Reflection in Engineering Education. (2014–2017). http://cpree.uw.edu/

Di Stefano, G. F., Gino, F., Pisano, G., & Staats, B. (2014). Learning by thinking: How reflection aids performance. *Harvard Business School Working Paper*, No. 14–093. http://www.sc.edu/uscconnect/doc/Learning%20by%20Thinking,%20How%20Reflection%20Aids%20Performance.pdf

Emerson, C., & Reid, A. (2019). Integrative learning and ePortfolio networks. In K. B. Yancey (Ed.), *ePortfolio as curriculum: Models and practices for developing students' ePortfolio literacy* (pp. 203–235). Stylus.

Flash, P. (2016). From apprised to revised: Faculty in the disciplines change what they never knew they knew. In K. B. Yancey (Ed.), *A rhetoric of reflection* (pp. 227–249). Utah State University Press.

Gawande, A. (2003). *Complications*. Picador.

Inoue, A., & Richmond, T. (2016). Theorizing the reflection practices of female Hmong college students: Is reflection a racialized discourse? In K. B. Yancey (Ed.), *A rhetoric of reflection* (pp. 126–146). Utah State University Press.

Kalman, C. S., Sobhanzadeh, M., Thompson, R., Ibrahim, A., & Wang, X. (2015). Combination of interventions can change students' epistemological beliefs. *Phys. Rev. ST Phys. Educ. Res., 11*, 020136–1–020136–17.

Lindemann, H., Camper, M., Jacoby, L. D., & Enoch, J. (2018). Revision and reflection: A study of (dis)connections between writing knowledge and writing practice. *College Composition and Communication, 69*(4), 581–611.

Neal, M. R. (2016). The perils of standing alone: Reflective writing in relationship to other texts. In K.B. Yancey (Ed.), *A rhetoric of reflection* (pp. 64–83). Utah State University Press.

Perl, S. (1980). Understanding composing. *College Composition and Communication, 31*(4), 363–370.

Pianko, S. (1979). Reflection: A critical component of the composing process. *College Composition and Communication, 30*(3), 275–278.

Rodgers, C. (2002). Defining reflection: Another look at John Dewey and reflective thinking. *Teachers College Record, 104*(4), 842–866.

Roozen, K. (2016). Reflective interviewing: Methodological moves for tracing tacit knowledge and challenging tacit chronotopic representations. In K. B. Yancey (Ed.), *A rhetoric of reflection* (pp. 250–268). Utah State University Press.

San Jose State University's "Reflection on college writing FAQ." (2019). https://www.sjsu.edu/english/frosh/how-to-choose/faq/index.php

Schön, D. (1987). *Educating the reflective practitioner: Toward a new design for teaching and learning in the professions.* Jossey-Bass.

Schrand, T., Jones, K., & Hanson, V. (2018). Reflecting on reflections: Curating ePortfolios for integrative learning and identity development in a general education senior capstone. *International Journal of ePortfolio, 8*(1), 1–12.

Silver, N. (2016). Reflection in digital spaces: Publication, conversation, collaboration. In K. B. Yancey (Ed.), *A rhetoric of reflection* (pp. 166–200). Utah State University Press.

Sommers, J. (2011). Reflection revisited: The class collage. *Journal of Basic Writing, 30*(1), 99–129.

Sommers, J. (2016). Problematizing reflection: Conflicted motives in the Writer's Memo. In K. B. Yancey (Ed.), *A rhetoric of reflection* (pp. 271–287). Utah State University Press.

Yancey, K. B. (1998). *Reflection in the writing classroom*. Utah State University Press.

Yancey, K. B., Robertson, L., & Taczak, K. (2014). *Writing across contexts: Transfer, composition, and sites of writing*. Utah State University Press.

Yancey, K. B. (2016 a). Introduction: Contextualizing reflection. In K. B. Yancey (Ed.), *A rhetoric of reflection* (pp. 3–20). Utah State University Press.

Yancey, K. B. (2016 b.) Defining reflection: The rhetorical nature and qualities of reflection. In K. B. Yancey (Ed.), *A rhetoric of reflection* (pp. 303–320). Utah State University Press.

Yancey, K. B. (2019). Creating an ePortfolio studio experience: The role of curation, design, and peer review in shaping ePortfolios. In K. B. Yancey (Ed.), *ePortfolio as curriculum: Models and practices for developing students' ePortfolio literacy* (135–149). Stylus.

Recommended Resources

Calvo, M. (2017). Reflective drawing as a tool for reflection in design research. *International Journal of Art & Design Education, 36*(3), 261–272. https://doi.org/10.1111/jade.12161

Kozubaev, S., Elsden, C., Howell, N., Søndergaard, M. L. J., Merrill, N., Schulte, B., & Wong, R. Y. (2020). Expanding modes of reflection in design futuring. In *Proceedings of the 2020 CHI Conference on Human Factors in Computing Systems* (pp. 1–15). ACM. https://doi.org/10.1145/3313831.3376526

McKenney, S., & Reeves, T. (2014). Methods of evaluation and reflection in design research. *Zeitschrift* für *Berufs- und Wirtschaftspädagogik, 27*, 141–153.

Quayle, M., & Paterson, D. (1989). Techniques for encouraging reflection in design. *Journal of Architectural Education, 42*(2), 30–42.

Suchman, L.A., & Trigg, R. H. (2020). Understanding practice: Video as a medium for reflection and design. In J. Greenbaum and M. Kyng (Eds.), *Design at work: Cooperative design of computer systems* (pp. 65–89). CRC Press.

See also: Memory, Heuristics, Cognitive Dissonance, Growth vs. Fixed Mindset

96. Repair

Thomas Karches

Repair is a process that returns a malfunctioning or non-functioning item to a working state. Replacement becomes more likely as the cost to repair the item approaches the cost of replacement. When items have sentimental value, repair can be more desirable, regardless of the cost. Repairing an item of higher quality is often preferable to replacing one with the same function but lesser quality.

Fig. 96.1. A user repairing a Mac computer. Photo by Revendo on Unsplash. https://bit.ly/keyword-fig-96–1

The ability to repair an item depends on the quality of the item, how it is constructed and the availability of replacement parts. Cheaply made items are often more difficult to repair because the repair process can cause additional damage. Items that are glued together are generally more difficult to repair than items that are manufactured using screws or other fasteners.

Repairing a broken device often requires parts of it to be replaced, which are usually available from the original manufacturer. Once a manufacturer stops making these replacement parts, the repair becomes more difficult. Replacement parts can often be salvaged from other broken devices of the same type and are sold locally and online. This option reduces the cost of parts and often allows a specific broken part to be purchased rather than an entire assembly of parts. For computers, a missing key from a laptop keyboard can be purchased online, where the manufacturer might only sell a replacement keyboard. Used automotive parts are available locally through "pick a part" companies that provide access to "junkyards" with vehicles from which parts can be salvaged.

Companies will often have their own "in house" repair organization. Repair by independent repair organizations is often discouraged and in some cases made very difficult due to specialized equipment that is required to perform repairs, which is often only available to repair facilities owned by the company. To replace certain parts in a John Deere tractor, the replacement parts must not only be mechanically compatible, but special software available only to authorized dealers is required to "reprogram" the parts to work with the rest of the tractor (Koebler, 2017). As described by farmer Kevin Kenney in Jason Koebler's (2017) report, "Deere charges $230, plus $130 an hour for a technician to drive out and plug a connector into their USB port to authorize the part" (n.p.). Some John Deere customers use "cracked" versions of this software to "reprogram" the replacement parts themselves. The introduction of "right to repair" laws has been the response to these repair restrictions, which legally provide access to these necessary parts and tools.

Repair is often the only option for devices that serve a unique need in a company. *The New York Times* printing press requires a three-person staff to keep it running. There is no option to quickly replace a device that occupies 840,000 square feet.

Evidence of a repair is usually hidden from view, as in a vehicle that is repaired at an automotive body shop. It is usually preferred that the repaired device looks the same as before. Alternatively, Kintsugi is the Japanese art of repairing pottery that uses gold or other metals to fill the cracks and bond the pieces together. Rather than attempting to hide the repair, it is highlighted. The damage becomes part of the history of the object.

References

Koebler, J. (2017, March 21). Why American farmers are hacking their tractors with Ukrainian firmware. *Vice.* https://www.vice.com/en_us/ article/xykkkd/why-american-farmers-are-hacking-their-tractors-with-ukrainian-firmware

Recommended Resources

Cangiano, S., Fornari, D., & Seratoni, A. (2019). Reprogrammed art, a bridge between the history of interactive art and maker culture. In V. Bradbury and S. O'Hara (Eds.), *Art hack practice: Critical intersections of art, innovation, and the maker movement* (pp. 31–39). Routledge.

Crosby, A., & Stein, J. A. (2020). Repair. *Environmental Humanities, 12*(1), 179–185. https://doi.org/10.1215/22011919–8142275

Doctorow, C. (2018, October 20). It's repair day: No one should be punished for "contempt of business model." Electronic Frontier Foundation. www.eff.org/deeplinks/2018/10/its-repair-day-no-one-should-be-punished-contempt-business-model.

Manandhar, S. (2018). Maker Culture in Nepal: Making vs. making. In E. Garber, L. Hochtritt, and M. Sharma (Eds.), *Makers, crafters, educators: Working for cultural change* (pp. 139–141). Routledge.

Oropallo, G. (2019). The fixing I: Repair as prefigurative politics. In G. Julier, A. Munch, M. N. Folkmann, H.-C. Jensen, and N. P. Skou (Eds.), *Design culture: Objects and approaches* (pp. 157–70). Bloomsbury.

Seravalli, A. (2017). *ReTuren: Participatory design, co-production and makers' culture for sustainable waste handling.* Doctoral dissertation. Malmö University.

See also: Maker Culture, DIY, Teardown, Tinkering, Perseverance

97. Teardown

Jason Markins

In the introduction to his 2004 "Maker's Bill of Rights," Peter "Mr. Ja-lopy" Vermeren wrote the oft quoted line, "If you can't open it, you don't own it." This motto reflects the popular sentiment within making and design communities that users have the right to alter, tinker, and hack devices they purchase, and it reflects the hands-on pedagogical approach of both as well. If you want to learn about something, you ought to take it apart and see for yourself how it works.

Simply put, a teardown is an act of taking something apart to learn how it works, what it is made of, or how one might alter it. Often the term is used to refer to educational activities that invite people to learn about their own electronic devices and to resist a consumer culture that throws away, trades in, or replaces a device when it breaks. At its most basic level, teardowns are responses to manufacturing companies that encourage these practices and discourage consumers from repairing, tinkering, or hacking their devices. Examples of different approaches to teardowns include the iF-ixit teardown of an Apple Watch (2018), the teardown of a network router (Mitchell, 2017), or the video of Matt Dawe taking apart a 1974 Harley Davidson (Davies, 2018).

In addition, the term "teardown" may refer to a technical document that labels a device's components and outlines how to take the device apart. A number of websites collect teardowns—the most notable being iFix-it, a website featuring community-generated repair manuals, teardown documents, and advice on working with tools and electronics. As a genre, these teardowns exist in large part as a response to the intentional lack of information provided by companies about their products. The genre itself is a rhetorical argument for consumer rights and an argument against the environmental and financial burdens of a throw-away culture regarding technology.

Teardowns can also refer to public events, such as when a lab, makerspace, or hackerspace invites community members to take part in events to tinker with various electronics. Often, organizations will host these teardowns under a moniker such as "Teardown Tuesdays" or "Breakerspace" (Makerspace for Education, n.d.), and the teardown events are used as a way to introduce children and adults to working with tools and technologies associated with making and design thinking. Or, a teardown may be an act of taking apart a device to learn how it is made by a customer, a journalist, or a competitor. Teardowns that are conducted to learn information about a product, such as when journalists perform a teardown on the latest apple product, can be contentious as corporations work to both obfuscate what consumers are capable of repairing, altering, or adjust on their own and to protect their intellectual property from competing manufacturers.

Fig. 97.1. Components of a digital camera. Photo by Vadim Sherbakov on Unsplash. https://bit.ly/keyword-fig-97–1

For this reason, teardowns are often considered ways of teaching the public about tools and technologies while also bringing into discussion questions and concerns about open access, consumer rights, and agency in relation to the electronic devices they use.

References

Apple Watch Series 4 teardown. (2018, Sept 24). iFixIt. https://www.ifixit.com/Teardown/Apple+Watch+Series+4+Teardown/113044

Davies, A. (2018, Nov 5). Luxuriate in this mechanic's teardown of a 1974 Harley Davidson. *Wired.* https://www.wired.com/story/harley-davidson-motorcycle-deconstructed-video/

Makerspace for Education. (n.d.). Breaker space—what is it? http://www.makerspaceforeducation.com/breaker-space.html

Mitchell, R. (2017, May 23). Teardown Tuesday: Mini network router. (2017). All About Circuits. https://www.allaboutcircuits.com/news/teardown-tuesday-mini-network-router/

Vemeren, P. (2004). Maker's bill of rights. *Make:*. https://makezine.com/article/maker-news/the-makers-bill-of-rights/

Recommended Resources

Brenner, D. (2015). Making by breaking: Why taking things apart is essential to making them work. *EdSurge.com*. https://www.edsurge.com/news/2015-05-21-making-by-breaking-why-taking-things-apart-is-essential-to-making-them-work

Collins, N. (2004). *Hardware hacking.* http://www.nicolascollins.com/texts/originalhackingmanual.pdf

Graham, S., & Thrift, N. (2007). Out of order: Understanding repair and maintenance. *Theory, Culture, and Society, 24*(3), 1–25.

iFixit. (n.d.). Repair manifesto. www.ifixit.com/Manifesto

McLellan, T. (2013). *Things come apart: A teardown manual for modern living.* Thames & Hudson.

Thorpe, A. (2010). Design's role in sustainable consumption. *Design Issues, 26*(2), 3–16.

van Dooren, T. (2014). Care. *Environmental Humanities, 4*(1), 291–294.

Willis, A.-M. (2014). Renew, repair, research. *Design Philosophy Papers, 9*(1), 1–3.

See also: Maker Culture, DIY, DIWO, Tinkering, Repair, Tools

98. Tinkering

Danielle Koupf

The terms "making" and "tinkering" are often tethered to each other, their distinctions collapsed. Mitchel Resnick and Eric Rosenbaum (2013) have disentangled them: tinkering is an open-ended approach to making that contrasts with following step-by-step instructions. In their words, "The tinkering approach is characterized by a playful, experimental, iterative style of engagement, in which makers are continually reassessing their goals, exploring new paths, and imagining new possibilities" (Resnick & Rosenbaum, 2013, p. 164). In contrast to planning, tinkering is messy and unpredictable and may begin without a goal, making it a bottom-up, rather than top-down, approach. Tinkering involves becoming stuck, persevering through the process, and finding ways forward through small breakthroughs (Bevan et al., 2015). Tinkers proceed through trial and error, relying upon the immediate feedback that a device, computer program, or collaborator provides. Tinkers may be autodidacts who learn on their own through the process of trying things out (Ito et al., 2010, p. 58).

Akin to *bricolage*, tinkering involves making do with everyday materials. But importantly, tinkering need not involve building something with physical resources: "We see tinkering as a style of making things, regardless of whether the things are physical or virtual. You can tinker when you are programming an animation or writing a story, not just when you are making something physical. The key issue is the style of interaction, not the media or materials being used" (Resnick & Rosenbaum, 2013, p. 166).

In fact, Jentery Sayers (2011) and Danielle Koupf (2017) have each applied the tinkering mindset (Martinez & Stager, 2013) to composition, language, and literature pedagogy. Sayers (2011) emphasizes the collaborative component of tinkering and the opportunities it provides for importing digital technologies into English studies. Drawing on the significance of iteration in tinkering, Sayers argues that tinkering helps students realize that writing and learning are nonlinear and that in fact an earlier iteration of an idea may actually be superior to a later iteration (p. 287). Koupf (2017) describes tinkering as a critical and creative approach to reusing and playing with existing texts, one that yields enhanced understanding of the source texts and new techniques for writing. Given early definitions of the *tinker* as an itinerant tinsmith, Koupf argues for the transferability of tinkering, suggesting that it is a skill students can take with them from one class or writing occasion to the next.

Tinkering has affinities with hacking, too (Varela, 2016; Wilkinson & Petrich, 2014, p. 6). Varela (2016) provides an example of tinkering, more than hacking, when he describes his approach to learning to code: "I was able to learn what I needed by watching online tutorials, asking around, failing over and over again, and thinking aloud through my mistakes" (p. 75). Varela made do with the available resources (online tutorials, asking around), proceeded through trial and error, and used immediate feedback to achieve desired results.

References

Bevan, B., Gutwill, J., Petrich, M., & Wilkinson, K. (2015). Learning through STEM-rich tinkering: Findings from a jointly negotiated research project taken up in practice. *Science Education, 99*(1), 98–120.

Ito, M., Baumer, S., Bittani, M., boyd, d., Cody, R., Herr-Stephenson, B., Horst, H. A., Lange, P. G., Mahendran, D., Martínez, K. Z., Pascoe, C. J., Perkel, D., Robinson, L., Sims, C., & Tripp, L. (2010). *Hanging out, messing around, and geeking out : Kids living and learning with new media.* MIT Press.https://clalliance.org/wp-content/uploads/files/Hanging_Out.pdf Koupf, D. (2017). Proliferating textual possibilities: Toward pedagogies of critical-creative tinkering. *Composition Forum, 35.* https://compositionforum.com/issue/35/proliferating.php.

Martinez, S., & Stager, G. (2013). *Invent to learn: Making, tinkering, and engineering in the classroom.* Constructing Modern Knowledge Press.

Resnick, M., & Rosenbaum, E. (2013). Designing for tinkerability. In M. Honey & D. Kanter (Eds.), *Design, make, play: Growing the next generation of STEM innovators* (pp. 163–181). Routledge.

Sayers, J. (2011). Tinker-centric pedagogy in literature and language classrooms. In L. McGrath (Ed.), *Collaborative approaches to the digital in English studies* (pp. 279–300). Utah State UP/Computers and Composition Digital Press.

Varela, M. (2016). Hacking and rehearsing: Experiments in creative tinkering. *New Theatre Quarterly, 32*(1), 68–77.

Wilkinson, K., & Petrich, M. (2014). *The art of tinkering.* Weldon Owen.

Recommended Resources

Cortez, J. (2009). Try it and fail [Video]. *Tinkering as a mode of knowledge production in a digital age. Carnegie views: Insights from education innovators.* https://vimeo.com/2225130

DiGiacomo, D., & Gutierrez, K. (2016). Relational equity as a design tool within making and tinkering activities. *Mind, Culture, & Activity*, 23(2), 141–153. https://doi.org/10.1080/10749039.2015.1058398

Franz, K. (2005). *Tinkering: Consumers reinvent the early automobile*. University of Pennsylvania Press.

Gross, M., & Do, E. (2009). Educating the new makers: Cross-disciplinary creativity. *Leonardo*, 42(3), 210–215.

Gutwill, J. P., Hido, N., & Sindorf, L. (2015). Research to practice: Observing learning in tinkering activities. *Curator: The Museum Journal*, 58(2), 151–168.

Vossoughi, S., & Bevan, B. (2014). *Making and tinkering: A review of the literature*. National Research Center. http://sites.nationalacademies.org/cs/groups/dbassesite/documents/webpage/dbasse_089888.pdf

Washor, E., & Mojkowski, C. (2010). *Making their way: Creating a new generation of "Thinkerers."* http://www.handoc.com/documents/makersymposium.pdf

See also: Creative Confidence, Play/Playful, Constructionism, FabLab, Makerspace

99. User stories

Ann Shivers-McNair

User stories are simplified descriptions of a user's requirements for a product that are written from the user's point of view and that articulate users' desires, goals, behaviors, and experiences. They are often associated with agile approaches to software development, which emphasize shorter development cycles that incorporate testing and implement changes, rather than a longer development cycle with the entire product implemented at the end (Cohn, 2004). But user stories are also applicable in a range of industries and applications, including public health (Turner, Reeder, & Ramey, 2013). Regardless of the industry or product, user stories can help designers and makers empathize with users and achieve clarity about user experiences and about the product.

Mike Cohn (2004) notes that user stories involve more than the written story itself: they also involve "conversations about the story that serve to flesh out the details of the story" and "tests that convey and document details and that can be used to determine when a story is complete" (p. 4). In this sense, user stories can serve as a dynamic accountability measure that helps a design team or maker articulate important findings from user research—through observations, interviews, surveys, the development of user personas, and other methods—and make sure the understandings of users' requirements are accurate.

Once a design team or maker has gathered an understanding of users' expectations and needs, they can write user stories—either from the perspective of specific individuals or from the perspective of user personas, or user archetypes drawn from user research—that connect users' requirements to a specific functionality in the product. Ideally, a user story corresponds to a single, specific functionality, so designers and makers may need to generate several user stories to address a range of requirements and functions. To this end, templates can be helpful: Garm Lucassen et al. (2015) recommend that a user story have a role, a means, and an end, as in the following:

> Role: As a Visitor
> Means: I want to register at the site
> End: so that I can contribute. (p. 7).

For example, a maker who is prototyping a smart watering system for plants might write a user story that articulates a common theme that emerged in conversations with potential users: as someone operating a smart watering system in my home, I want to see how much water the system uses so I can plan for my water bill.

User stories are often abstractions of what multiple and different users say and do, and the template structure and focus on simplicity can help designers and makers streamline user research data in a way that is easy to remember and incorporate in the design and testing process. But while abstraction and simplification can be useful, Natasha N. Jones (2016) offers an important counterpoint in favor of preserving the complexity of what users say: "Use actual user narratives, users' own stories in their own voices, to replace traditional scenarios, avoiding the decontextualization of user experiences for design purposes and embracing and valuing the knowledge that a user can provide about a design" (p. 486). This is a helpful reminder to continually and actively de-center the designer's perspective, either in the user stories themselves or in pairing user stories with other methods of centering users' voices.

References

Cohn, M. (2004). User stories applied: For agile software development. Pearson Education Inc.

Jones, N. N. (2016). Narrative inquiry in human-centered design: Examining silence and voice to promote social justice in design scenarios. *Journal of Technical Writing and Communication, 46*(4), 471–492.

Lucassen, G., Dalpiaz, F., van der Werf, J. M. E., & Brinkkemper, S. (2015, August). Forging high-quality user stories: towards a discipline for agile requirements. In *2015 IEEE 23rd International Requirements Engineering Conference* (pp. 126–135). IEEE.

Turner, A. M., Reeder, B., & Ramey, J. (2013). Scenarios, personas and user stories: User-centered evidence-based design representations of communicable disease investigations. *Journal of Biomedical Informatics, 46*(4), 575–584.

Recommended Resources

Amna, A. R., & Poels, G. (2022). Ambiguity in user stories: A systematic literature review. *Information and Software Technology, 145,* 106824. https://doi.org/10.1016/j.infsof.2022.106824

Desharnais, J. M., Buglione, L., & Kocatürk, B. (2011, June). Using the COSMIC method to estimate Agile user stories. In *Proceedings of the 12th international conference on product focused software development and process improvement* (pp. 68–73). ACM. https://doi.org/10.1145/2181101.2181117

Smith, L. T. (2019). *Decolonizing research: Indigenous storywork as methodology*. Bloomsbury Publishing.

Wautelet, Y., Heng, S., Kiv, S., & Kolp, M. (2017). User-story driven development of multi-agent systems: A process fragment for agile methods. *Computer Languages, Systems & Structures, 50*, 159–176. https://doi.org/10.1016/j.cl.2017.06.007

Zeaaraoui, A., Bougroun, Z., Belkasmi, M. G., & Bouchentouf, T. (2013). User stories template for object-oriented applications. In *Third International Conference on Innovative Computing Technology (INTECH 2013)* (pp. 407–410). IEEE. doi:10.1109/intech.2013.6653681

See also: User Requirements, User Experience, Iterative Design

100. Wireframing

Joseph Bartolotta

One of the early steps in designing an interactive system is composing a "wireframe" model or diagram. A wireframe is a simple drawing that shows the general layout of the interactive system. It shows where different components of content might belong, including text, images, links, search bars, navigation, and logos, to name a few. Wireframes are most useful for designers before they develop prototypes so they can get a sense of how visual elements will take up space in the interactive system. For example, Figures 100.1 and 100.2 show how a mobile application for Delta Airlines may be represented as a wireframe. Keep in mind that while developing wireframes based on existing interactive systems may be a useful tool, the practice in user-centered design is to compose a wireframe for the purpose of seeing how important design elements will appear on a page. Designers should think about the user while creating a wireframe.

Good wireframes focus on content placement rather than visual design (or user interface design, or UI). Designers should not be spending time trying to make the wireframe look beautiful as it may slow down the iteration process or introduce unnecessary confusion to the user during testing. An effective wireframe should allow designers to observe user journey--i.e., the path that users take to complete a task or action on the interface, like purchasing an item or downloading a file. As such, wireframes typically include multiple pages to represent the different screens a user may encounter during a task.

Wireframes help designers divvy up what will take up space in the interactive system. They give designers an opportunity to break the design process into more manageable chunks and focus the design process.

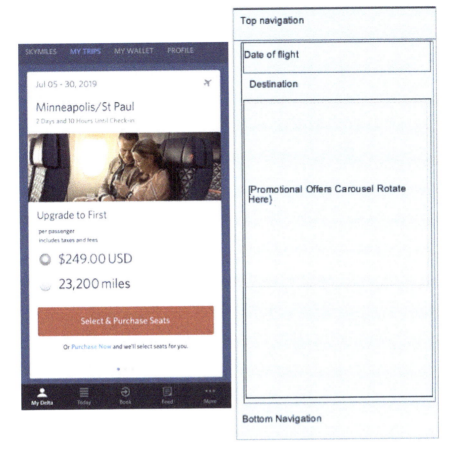

Fig. 100.1. Delta Airlines app homepage. Screenshot by the author.
Fig. 100.2. Delta Airlines app homepage wireframe. Screenshot by the author.

Recommended Resources

Buxton, B. (2010). *Sketching user experiences: Getting the design right and the right design*. Morgan Kaufmann.

de Lange, P., Nicolaescu, P., Neumann, A. T., & Klamma, R. (2020). Integrating web-based collaborative live editing and wireframing into a model-driven web engineering process. *Data Science and Engineering, 5*, 240–260. https://link.springer.com/article/10.1007/s41019-020-00131-3

Garrett, J. J. (2011). *The elements of user experience: User-centered design for the web and beyond (2nd Ed.)*. New Riders.

Grewenig, S. (2013). *From high-usability cross-device wireframe-based storyboards to component-oriented responsive-design user interfaces.* Doctoral dissertation. Augsburg University. http://architecturedissertation.com/publication/MasterThesis-Grewenig.pdf

Hamm, M. J. (2014). *Wireframing essentials.* Packt Publishing Ltd.

Robbins, J. N. (2012). *Learning web design: A beginner's guide to HTML, CSS, JavaScript, and web graphics (4th Ed.).* O'Reilly Media, Inc. p. 6.

See also: Interaction Design, Iterative Design, Visual Rhetoric

Contributors

Kamila Albert (PhD, Florida State University) is the director of Florida State University's reading-writing center and digital studio. Kamila's research interests include multimodal composition, design thinking, digital and visual literacy, and writing center studies.

Erin Kathleen Bahl (PhD, The Ohio State University) is an associate professor of applied and professional writing at Kennesaw State University in Atlanta, Georgia. She is managing editor for *Kairos: A Journal of Rhetoric, Technology, and Pedagogy*. Her work explores the possibilities digital technologies afford for creating knowledge and telling stories, with a focus especially on webtexts and webcomics. Her publications include work in *Kairos*; *Computers and Composition Online*; Computers and Composition Digital Press; *enculturation: a journal of rhetoric, writing, and culture*; *The Digital Review*; *The Journal of American Folklore*; and the Smithsonian's *Folklife Magazine*.

Sweta Baniya (PhD, Purdue University) is an assistant professor of rhetoric and professional and technical writing and an affiliate faculty of the Center for Coastal Studies and Women and Gender Studies at Virginia Polytechnic Institute and State University. Through a transnational and non-Western perspective, her research focuses on transnational coalitions in disaster response, crisis communication, and non-Western rhetorics. Her forthcoming book *Transnational Assemblages: Social Justice Oriented Technical Communication in Global Disaster Management* explores transnational activism in the April 2015 Nepal Earthquake and 2017 Hurricane Maria in Puerto Rico.

Joseph Bartolotta (PhD, University of Minnesota) is an associate professor in writing studies and rhetoric at Hofstra University.

Estee Beck (PhD, Bowling Green State University) works as the Merritt Writing Program director in the School of Social Sciences, Humanities, and Arts at the University of California–Merced.

Arthur Berger works as a technical writer on open source software and cloud technologies at Solo.io, an application networking startup. His professional interests include technical communication, content strategy, design thinking, and developer experience (DX), especially application program interface (API) documents, user interface (UI) text, command line interface (CLI) strings, and error messages. He has a MS degree in technical communication from North Carolina State University and is an active member of the Society for Technical Communication Carolina chapter.

Elin A. Björling (PhD, University of Washington) is a senior research scientist and affiliate faculty member in Human-Centered Design and Engineering at the University of Washington–Seattle. Over the past two decades, she has studied adolescent health utilizing a mixed-methods approach with a focus on momentary data. Most recently she has engaged adolescents through participatory design in the development of new technologies for mental health.

Casey Boyle (PhD, University of South Carolina) is an associate professor in the Department of Rhetoric and Writing at the University of Texas at Austin and director of the Digital Writing and Research Lab, where he researches and teaches digital rhetoric, media studies, and/as rhetorical history. His work has appeared in *Kairos, Philosophy and Rhetoric, Computers and Composition, Technical Communication Quarterly, College English* as well as essay collections. His book, *Rhetoric as a Posthuman Practice*, explores the role of practice and ethics in digital rhetoric.

Kevin Brock (PhD, North Carolina State University) is an associate professor in the Department of English Language and Literature at the University of South Carolina, where he teaches courses on professional writing, composition, and digital rhetoric. He is the author of *Rhetorical Code Studies: Discovering Arguments in and around Code* (U of Michigan P).

Emily F. Brooks (PhD, University of Florida) specializes in book history, children's literature and culture, and digital humanities. She has written several library guides and taught Arduino courses for middle schoolers at Girls Tech Camp, high schoolers at Gator Computing Camp, and university students at the Marston Science Library. She regularly incorporates teaching physical computing and digital fabrication in her courses.

Antonio Byrd (PhD, University of Wisconsin–Madison) is an assistant professor of English at the University of Missouri–Kansas City where he teaches courses in Black literacies, professional and technical communication, multimodal writing and rhetorics, digital rhetoric, and composition. His research studies how the legacies of liberation carry forward into Black digital literacies, with a special focus on Black adults learning and using computer programming. His work has appeared in *College Composition and Communication, Literacy in Composition Studies, Technical Communication Quarterly*, and *Writer: Craft and Context*. He is the winner of the 2021 CCCCs Braddock Award. His current book project is *The Literacy Pivot: How Black Adults Learn Computer Programming in a Racist World*, which is under contract with the WAC Clearinghouse/University of Colorado Press.

Erin Brock Carlson (PhD, Purdue University) is an assistant professor in the Department of English at West Virginia University, where she teaches graduate and undergraduate courses in the Writing Studies program. Her current research addresses community organizing, place-based development, and collaborative knowledge making through participatory visual methodologies. Her work has appeared in *Computers and Composition*, *Kairos*, *Journal of Business and Technical Communication*, *Technical Communication Quarterly*, and several edited collections.

Alexandra Catá-Ross is a PhD candidate in North Carolina State University's Communication, Rhetoric, and Digital Media program focused on intersections between technical communication and video game studies. Her dissertation research focuses on technical communication work in video game development environments, and is currently an information architect at Epic Games. She has several publications across esports, game-based learning, and chatbot design in various places, including *Technical Communication* and *Journal of Business and Technical Communication*.

Mary E. Caulfield (A.L.M., Harvard University) is a lecturer in MIT's Writing, Rhetoric, and Professional Communications department. She works with university-level students on project-based classes in design and engineering and teaches critical thinking and research skills to advanced secondary-school students. She has spoken at conferences on writing, speaking, and teamwork and has moderated panels on project-based learning, youth, and the media. Prior to joining MIT, she was a writer in the consulting and software industries.

Dennis Cheatham (MFA, University of North Texas) is an associate professor of communication design at Miami University in Oxford, Ohio. He is graduate director of the MFA program in Experience Design and a research fellow in Miami's Scripps Gerontology Center. Dennis's research examines the interplay between design and people at an experiential level. His current work focuses on end-of-life choices and topics in aging, specifically advance care planning. Dennis practiced design for fifteen years in Dallas/Fort Worth, Texas.

Joe Cirio (PhD, Florida State University) is assistant professor of writing and first-year studies and convenor for the WAC program at Stockton University in Galloway, New Jersey. He teaches undergraduate courses on the making of memory, professional writing and design, and first-year writing. He also serves as co-coordinator for the Digital Praxis Posters at CCCC. He has published and presented on topics related to everyday writing, community-based assessment practices, and community literacy.

Jess Clements (PhD, Purdue University) is associate professor of English and director of the Composition Commons at Whitworth University. She currently serves as managing editor for *Present Tense* (https://www.presenttensejournal.org/). Her scholarship centers on *ethos* and the role of human and object-oriented actors in contemporary multimodal communication. She collaborated on an interdisciplinary book, *Optimal Motherhood and Other Lies Facebook Told Us*, evaluating the influence of social media networks in shaping binary-bound parenting decisions (https://mitpress.mit.edu/9780262543620/optimal-motherhood-and-other-lies-facebook-told-us/). She has published on the pedagogical performance of faith in the first-year writing classroom as well as writing center tutor training topics such as the role of new media expertise in shaping writing center consultations and gaming ethnography as a pathway to raising intersectional awareness among tutors. Whitworth University's Composition Commons website: https://www.whitworth.edu/compositioncommons/

Justin Cook (PhD, Texas Woman's University) is an assistant professor of English and director of the Writing Center at High Point University. His work focuses on the material-semiotic discursivity of digitized, politicized, and stigmatized bodies. He also centers queer methods and multimodality in his research praxis, specifically in regard to teaching technical writing and teaching in digital learning environments.

Jacob Craig (PhD, Florida State University) is an associate professor of English at College of Charleston where he teaches courses in digital rhetoric, composition theory, and technical writing. His research examines the relationships between writers and their material worlds: particularly, writers' technologies and their locations of writing. His eportfolio is available at http://jacobwcraig.com.

Sara Doan (PhD, University of Wisconsin-Milwaukee) is an assistant professor of experience architecture at Michigan State University. Her research examines how expertise is framed and enacted across different genres, such as instructor feedback on resumes and cover letters, misleading data visualizations about COVID-19, audience co-creation in public service announcements, and the content strategy of Southeastern state health departments.

Dana Lynn Driscoll (PhD, Purdue University) is a professor of English and director of the Kathleen Jones White Writing Center at Indiana University of Pennsylvania, where she teaches in the Composition and Applied Linguistics doctoral program. She serves as co-editor for the open-source first-year writing textbook series *Writing Spaces*, which offers free readings and instructional materials for composition courses. She has published widely on writing transfer, learning theory, writing centers, and research methods and has offered plenary addresses and workshops around the globe.

Renee Ann Drouin (PhD, Bowling Green State University) is an assistant teaching professor of rhetoric and composition at Penn State Harrisburg. Her research interests include feminist rhetorics, digital ethnographies, and popular culture analysis. Current projects include applications of ancient feminist rhetorics to video games and ethical considerations in performing ethnographies across social media platforms.

Ann Hill Duin (PhD, University of Minnesota) is a professor of writing studies and Graduate-Professional Distinguished Teaching Professor at the University of Minnesota where her research and teaching focus on digital literacy, collaboration, and workplace writing futures. In 2021 she received both the 2021 Ronald S. Blicq Award for Distinction in Technical Communication from the IEEE Professional Communication Society and the 2021 J. R. Gould Award for Excellence in Teaching from the Society for Technical Communication. Together with Isabel Pedersen, see *Writing Futures: Collaborative, Algorithmic, Autonomous* (Springer, 2021, https://bit.ly/keyword-citation-wf) and *Augmentation Technologies and Artificial Intelligence in Technical Communication: Designing Ethical Futures* (Routledge, 2023, https://bit.ly/keyword-citation-at).

Sergio C. Figueiredo (PhD, Clemson University) is an associate professor of media and rhetoric in the Department of English at Kennesaw State University. He is the translator of *Inventing Comics: A New Translation of Rodolphe Töpffer's Essays on Graphic Storytelling, Media Rhetorics, and Aesthetic Practice.*

Merideth Garcia (PhD, University of Michigan) is an associate professor of English at the University of Wisconsin–La Crosse, where she teaches courses in composition, literature, and secondary ELA pedagogy. Her research investigates the social dynamics of networked classrooms, the literacy practices of fan communities, and the construction of inclusive and sustainable teaching and learning environments.

Thomas M. Geary (PhD, University of Maryland) is a professor of English at the Virginia Beach campus of Tidewater Community College, where he teaches composition, rhetoric, technical writing, developmental writing, and humanities courses. Tom currently serves as the editor of *Inquiry*, a peer-reviewed journal for faculty, staff, and administrators in the Virginia Community College System. His academic interests include sonic rhetoric, electracy, podcasting, digital storytelling, open educational resources, visual rhetoric, and compassionate pedagogy.

Krys Gollihue (PhD, North Carolina State University) is a writer, designer, and teacher specializing in technical communication, design thinking, open source technology, and sustainable agriculture. They currently support product and digital experiences at Red Hat, developing instructional and trial content for enterprise Linux and the open hybrid cloud.

Katherine Goodman (PhD, University of Colorado Boulder) is an associate professor in the College of Engineering, Design, and Computing, University of Colorado Denver, where she teaches courses in human-centered design. Her research seeks to understand the roles of creativity and collaboration in learning, and she is active in the American Society for Engineering Education. Her recent article in *Leonardo* is titled "Surely You Must Be Joking, Mr. Twain! Re-engaging Science Students Through Visual Aesthetics" (https://doi.org/10.1162/leon_a_01604).

Jacob Greene (PhD, University of Florida) is an assistant professor of English in the Writing, Rhetorics, and Literacies program at Arizona State University. His research explores the rhetorical potential of emerging location-based writing technologies, from mobile augmented reality applications to GPS-guided audio tours. His work has appeared in *Composition Studies*, *Enculturation*, *Computers & Composition Online*, and *Kairos*.

Lyra Hilliard (M.F.A., Utah State University) is a principal lecturer in the Department of English at the University of Maryland, where she coordinates the undergraduate teaching assistant program and the blended and online writing division of the academic writing program. She teaches academic, digital, and professional writing as well as the teaching practicum for undergraduate teaching assistants in English. She co-facilitates the mentoring program for blended and online teaching and helped design the department's Professional Track Faculty (PTK) Mentoring Program.

Steve Holmes (PhD, Clemson University) has been an associate professor of technical communication and rhetoric at Texas Tech University since 2019. His publication areas include digital rhetoric, ethics, technical communication and rhetorical theory. Steve is the author of *Procedural Habits* (2017) on the embodied rhetorics and videogames and *Rhetoric, Technology and the Virtues* (2018) on social media ethics (with Jared S. Colton). His scholarship on critical making includes a Fall 2022 edited collection with Michael J. Faris on critical making, which is entitled *Reprogrammable Rhetorics*.

Ada Hubrig (PhD, University of Nebraska–Lincoln) works as an assistant professor of English at Sam Houston State University, where they also serve as English education coordinator and co-chair of the composition program. Their research centers on disability, drawing from disability justice and queer theory to create anti-ableist and anti-oppressive pedagogies centering multiply-marginalized folx. Their writing appears in *College Composition and Communication, Teaching English in the Two-Year College, Pedagogy, Community Literacy Journal, Reflections, Journal of Multimodal Rhetorics, Present Tense, QED: A Journal in GLBTQ Worldmaking*, among other journals and edited collections.

Cody Jackson is a PhD candidate in rhetoric-composition at Texas Christian University in Fort Worth, Texas. Cody's research-pedagogy focuses on the intersections between disabled embodiment, queerness, multimodality, and archival theory-practice.

Jialei Jiang (PhD, Indiana University of Pennsylvania) is a teaching assistant professor of composition at the University of Pittsburgh. Her research interests include digital rhetoric, feminist posthumanism, composition studies, and multimodal pedagogy. Trained in qualitative research, she focuses on examining the use of emerging technologies and multimodal projects in college composition courses. She is also interested in exploring the intersections between multimodality, diversity, and pedagogy. Her works have appeared in *College Composition and Communication* (Forthcoming), *Composition Forum, Computers and Composition, Journal of Technical Writing and Communication, Kairos*, and edited collections.

Madison Jones (PhD, University of Florida) is an assistant professor of professional & public writing/natural resources science at the University of Rhode Island. Situated within the fields of environmental rhetoric, public advocacy, and place-based writing, his research focuses on how to use digital/visual technologies to better engage local communities when addressing large-scale environmental problems. His research combines place-based writing practices with locative media (such as augmented reality and digital maps) to understand the ways that technologies mediate environmental issues, foster more-than-human ethics, and shape a sense of place.

Thomas Karches (M.Ed., North Carolina State University) is an IT systems specialist and educator at North Carolina State University. He is working to raise awareness of the importance of repair and sustainability. He is actively involved in a local Repair Cafe chapter, which provides free repair services with an emphasis on demystifying the repair process.

JongHun Kim (PhD, University of Massachusetts Amherst) is the associate director of education at the Handel and Haydn Society.

Matthew Kim (PhD, Illinois State University) is a faculty fellow and the English department chair at Eagle Hill School in Hardwick, MA. He and Russell Carpenter are the editors of *Writing Studio Pedagogy: Space, Place, and Rhetoric in Collaborative Environments.*

Danielle Koupf (PhD, University of Pittsburgh) is an associate teaching professor in the Writing Program at Wake Forest University whose research focuses on invention, textual reuse, style, and rhetorics of making and crafting. Her publications have introduced the concepts of critical-creative tinkering and scrap writing to rhetoric and composition.

Ashanka Kumari (PhD, University of Louisville) is an assistant professor of English composition and rhetoric at Texas A&M University–Commerce. Her research focuses on first-generation-to-college students, graduate study, identity and literacy studies, multimodal and digital composing, and popular culture. Her work has appeared in *WPA Journal*, *Kairos: A Journal of Rhetoric, Technology, and Pedagogy*, *Composition Studies*, and the *Journal of Popular Culture*.

R.J. Lambert (PhD, University of Texas at El Paso) is a survivor of the Columbine High School shootings and studies risk communication, harm prevention, and crisis response related to medical, environmental, and institutional traumas. He previously coordinated cancer research at Harborview Medical Center and now teaches science writing and health communication in the Center for Academic Excellence and Writing Center at the Medical University of South Carolina.

Liz Lane (PhD, Purdue University) is an associate professor and coordinator of the Writing, Rhetoric, and Technical Communication program at the University of Memphis. She teaches courses such as document design, web design and online writing, and undergraduate and graduate seminars in technical writing. She has published widely on activism, digital rhetoric, and technical communication.

Halcyon M. Lawrence (PhD, Illinois Institute of Technology) is an associate professor of technical communication and information design at Towson University. She has over twenty years of professional experience as a technical trainer, writer, and usability practitioner. Her research focuses on speech intelligibility and the design of speech interactions for voice technologies, particularly for under-represented user populations.

Heather Listhartke is a PhD candidate at Miami University in composition and rhetoric. She studies the intersections of access and community within composing and making spaces. Her related research interests include digital literacy, technical and professional writing, and cultural rhetorics. Heather also enjoys working in writing curriculum development across disciplines.

Jason Markins (PhD, Syracuse University) is an adjunct instructor at Eckerd College.

Megan Marshall (PhD, University of Wyoming) is an associate professor of English at Marshall University, where she serves as the composition coordinator and teaches courses in writing pedagogy, embodied rhetoric, young adult literature, and composition. She has published and presented her work in the areas of young adult literature, multigenre composition, and the sociopolitical function of text-to-screen adaptation/s in popular culture.

Amanda M. May (PhD, Florida State University) is an assistant professor of English and director of the Writing Center at New Mexico Highlands University. In addition to teaching writing courses at the first-year, undergraduate, and graduate level, she supports NMHU students at the Writing Center on-site and online in synchronous and asynchronous sessions. She has additional experience in developing resources, including tip sheets, handouts, and workshops, for institutional use. Her recent research focuses on writing center history, writing center social media use and non-use, and aleatory play in video games.

Marijel (Maggie) Melo (PhD, University of Arizona) is an assistant professor in the School of Information and Library Science at the University of North Carolina at Chapel Hill. Her research resides at the intersection of innovation, critical maker culture, and the development of equitable collaborative learning spaces (e.g., makerspaces) in academic libraries. She co-founded the University of Arizona's first publicly accessible and interdisciplinary makerspace—iSpace—and strategically facilitated its growth from a four hundred-square-foot room in the Science-Engineering Library to a five thousand-square-foot facility.

Chloe Anna Milligan (PhD, University of Florida) is assistant professor of digital humanities in the Writing and Digital Media program at Pennsylvania State University, Berks College. She holds degrees in English from the University of Florida (PhD), Clemson University (MA), and Emmanuel

335

College (BA). She teaches, researches, and publishes widely on electronic literature, embodied rhetorics, game studies, media archaeology, and multimodal composition.

Kristen R. Moore (PhD, Purdue University) is an associate professor of technical communication in the Departments of Engineering Education and English at the University at Buffalo. Her research focuses on the public forms of technical communication, with a particular interest in how to increase the participation, efficacy, and justice of public technical communication. Her research has been published in a range of edited collections and many journals, including *Journal of Technical Writing and Communication, Technical Communication Quarterly*, and *IEEE Transactions on Professional Communication*. Most recently, she co-authored *Technical Communication After the Social Justice Turn: Building Coalitions for Action*.

Joe Moses (PhD, University of Minnesota) is senior lecturer in writing studies at the University of Minnesota–Twin Cities, where he teaches in the program in technical writing and communication and conducts professional development workshops on collaborative writing. He has published on design thinking, collaborative writing, project management, intercultural connectivism, and personal learning networks. He is co-author with Jason Tham of the *Collaborative Writing Playbook: An Instructor's Guide to Designing Writing Projects for Student Teams* (Parlor Press, 2021).

Derek N. Mueller (PhD, Syracuse University) is professor of rhetoric and writing at Virginia Tech. His teaching and research attends to the interplay among writing, rhetorics, and technologies. Mueller regularly teaches courses in visual rhetorics, composition theory, and research methods. He continues to be motivated professionally and intellectually by questions concerning digital writing platforms, networked writing practices, and discipliniographies or field narratives related to writing studies/rhetoric and composition.

Jeff Naftzinger (PhD, Florida State University) is assistant professor of rhetoric, composition and writing at Sacred Heart University in Fairfield, Connecticut. He teaches courses on digital writing and rhetoric, everyday writing, and courses in the first-year seminar. He has published and presented on topics related to defining and illustrating everyday writing, sustaining multimodal composing, and writing in digital spaces.

Mandy Olejnik (PhD, Miami University) serves as the assistant director of Writing Across the Curriculum at Miami University, where she supports disciplinary faculty across divisions through her work at the Howe Center for Writing Excellence.

Nitya Pandey is a PhD candidate in rhetoric and composition at Florida State University, Tallahassee. She is originally from Nepal. Since she is situated at the junction of different cultures, she is interested in all things intercultural. Moreover, she is intrigued by the concepts of multimodality, affect, and online writing instruction, and response. Whenever she is free, she likes to video chat with her dog, cook something edible, and take long, thoughtful walks.

A. Nicole Pfannenstiel (PhD, Arizona State University) is an assistant professor of digital media in the English department at Millersville University of Pennsylvania, where she teaches web writing, content management, first year composition, and new media literacy courses. She serves as the graduate coordinator for the MA and MED English programs. She also serves on the Open Education Committee, overseeing grants and administering programs to support faculty adoption of open access materials in their courses.

Megan Poole (PhD, Penn State University) is an assistant professor of English at the University of Louisville. Her interdisciplinary research and teaching centers on rhetorical theory, science studies, and technical writing, particularly focusing on how rhetorics of science can make science more accessible for public audiences.

Kristopher Purzycki (PhD, University of Wisconsin-Milwaukee) is an assistant professor of English and humanities at the University of Wisconsin–Green Bay who studies digital media, equitable pedagogy, and virtual placemaking. He is currently developing an original OER for first-year professional writers as well as researching the impacts of AI-driven content generators on creative practice.

Johansen Quijano (PhD, University of Texas at Arlington) is an associate professor of English at Tarrant County College Trinity River Campus who has been involved in maker culture since 2017 when he partnered with the Texas Makers Guild to create a prototype video game controller for a gaming and disability awareness event. He has since approached his teaching and research with a maker mindset. He has presented on maker culture at several conferences, schools, and colleges in the US.

Nupoor Ranade (PhD, North Carolina State University) is an assistant professor at George Mason University, where she teaches undergraduate and graduate courses on professional writing and rhetoric. Her research combines her technological background, computational skills, and her ability to address knowledge gaps in the fields of technical communication pedagogy and practice, and focuses on audience analysis, data analytics, user experience, information design in the fields of content production, software development, and artificial intelligence.

Cody Reimer (PhD, Purdue University) is an associate professor of English at the University of Wisconsin–Stout, where he teaches in the Professional Communication and Emerging Media undergraduate program and Technical and Professional Communication master's program. He serves on the editorial board for *Communication Design Quarterly*, the peer-reviewed research publication of the Association for Computing Machinery (ACM) Special Interest Group for Design of Communication (SIGDOC). His research explores the intersection of technical writing in videogames.

Max Renner (PhD, North Carolina State University) is an assistant professor at Molloy University. His research focuses on rhetorical invention, new media, and place. His publications examine how understandings and experiences of place, both materially and digitally, impact the development, performance, and identity of publics.

Michael Riendeau (Ed.D., University of Massachusetts) has directed the academic programs at Eagle Hill School since 1997. He continues to teach in the English department, where he began at Eagle Hill School in 1989 after completing his A.B. in English at the College of the Holy Cross. His professional and research interests include disability theory, multicultural education, and teacher preparation and induction.

Laura Roberts (PhD, North Carolina State University) is an assistant professor at University of Wisconsin–Platteville, where she teaches courses in technical writing, digital writing, and usability studies. Her research interests include rhetoric of science, public science communication, and technical communication.

Jennifer Sano-Franchini (PhD, Michigan State University) is Gaziano Family Legacy Professor of Rhetoric and Writing and associate professor of English at West Virginia University. Her scholarly interests are in interaction design, user experience, cultural rhetorics and/of technology, and Asian American rhetoric. She has published in journals including *Technical Communication*, *Enculturation*, *Rhetoric Review*, and *Rhetoric, Professional Communication, and Globalization*, as well as edited collections such as *Rhetoric and Experience Architecture* and *Rhetoric and the Digital Humanities*. Prior to going into academia, she spent seven years as the design consultant for a small copy company in Honolulu, Hawaiʻi.

Joy Santee (PhD, Purdue University) is an associate professor at the University of Southern Indiana where she teaches in the Professional Writing and Rhetoric program. Previously, she co-created the Writing Studies program at Utah Valley University and the Professional Writing and Rhetoric program at McKendree University. Her current research focuses on cartographic literacy in multimodal contexts, and she has been published in *Prompt: A Journal of Academic Writing Assignments, Journal of Technical Writing and Communication* and *College Teaching.*

Ann Shivers-McNair (PhD, University of Washington Seattle) is associate professor and director of professional and technical writing in the Department of English at the University of Arizona, where she also serves as co-organizer of UX@UA, a user experience community in Tucson. She publishes on the maker movement, qualitative research methods, and social justice-focused user experience. Current research projects include National Science Foundation-supported studies of equity-centered design learning in makerspaces and engineering contexts.

John Joseph Silvestro (PhD, Miami University) is an assistant professor of languages, literatures, cultures, and writing at Slippery Rock University, where he teaches courses in critical writing, grant writing, writing for advocacy, professional writing, writing for video games, and content writing. He draws upon qualitative research methods (interviews, surveys, content analysis) to study how writers compose texts that seek to generate local change and how writers' literacies have become entwined with algorithmic systems. He has published on circulation strategies, audience theories, rhetorical approaches to advocacy, and the history of spell-checker algorithms.

Scott Sundvall (PhD, University of Florida) is an assistant professor of English at the University of Memphis. His edited collection, *Rhetorical Speculations: The Future of Rhetoric, Writing, and Technology* (2019), uses a speculative approach for understanding the technological potential and future of rhetoric and writing. His work has also appeared in *Philosophy and Rhetoric, Composition Forum, Computers and Composition, Textshop Experiments,* and *Media Fields.*

Jason Tham (PhD, University of Minnesota) is an associate professor of technical communication and rhetoric at Texas Tech University, where he teaches user experience research, information design, discourse and technology, and instructional design. He is author of *Design Thinking in Technical Communication* (2021, Routledge) and editor of *Keywords in Design Thinking* (2022,

WAC Clearinghouse/University Press of Colorado). He currently serves as vice president of the Council for Programs in Technical and Scientific Communication. He is incoming editor of *Computers and Composition*.

Quentin D. Vieregge (PhD, University of South Florida) is a professor of English at the University of Wisconsin-Eau Claire–Barron County. He teaches first-year composition, advanced composition, business writing, literature, and film courses. He has published in the fields of rhetoric and composition and popular culture. He is interested in the scholarship of teaching and learning, collaborative writing, film analysis, and pop culture and politics.

Sarah Welsh (PhD, University of Texas at Austin) is a program manager at All Tech Is Human, where she focuses on mentorship in the responsible technology field. She also consults with AI and machine learning companies on strategic communications, public relations, and growth marketing. She lives in Austin, Texas.

Stephanie West-Puckett (PhD, East Carolina University) is an assistant professor of professional and public writing at the University of Rhode Island. She teaches cultural rhetorics and directs the first-year writing (FYW) program, which is heavily influenced by her research and participation in makerspaces. Her scholarship focuses on equity-enabled pedagogy and assessment and has been published in several edited collections and journals including *College English, Journal of Multimodal Rhetorics,* and *Community Literacy Journal.* Most recently, she co-authored *Failing Sideways: Queer Possibilities for Writing Assessment.*

Charles Woods (PhD, Illinois State University) is an assistant professor of English at Texas A&M University–Commerce, where he teaches undergraduate and graduate courses in rhetoric, composition, and technical communication. His research interests include digital rhetorics, privacy and surveillance studies, and podcasting. Out of his work in graduate school came the Digital Rhetorical Privacy Collective—an interactive, online resource for those interested in learning more about the intersections of digital privacy, policy, and power.

Kathleen Blake Yancey (PhD, Purdue University), Kellogg Hunt Professor and Distinguished Research Professor Emerita at Florida State University, has served as chair/president of several literacy organizations, as editor of *College Composition and Communication*, and as PI for three research projects investigating writing curricula supporting students' transfer of writing knowledge and practice. A member of several boards, she has edited/

authored sixteen books, often with colleagues, and over one hundred articles and book chapters; she has received several awards for her teaching, mentorship, and scholarship.

Sarah Young (PhD, Arizona State University) is a postdoctoral researcher in digitization at Erasmus University Rotterdam. She studies workplace privacy, surveillance, ethics, ICTs, and technical communication and has twelve years of experience doing national security and public trust background investigations in the United States. She is also a Center for Quantum Networks fellow researching the social impact of quantum networks.

www.ingramcontent.com/pod-product-compliance
Lightning Source LLC
Chambersburg PA
CBHW041639050326

40690CB00027B/5270